The Economics of Forestry and Natural Resources

The Economics of Forestry and Natural Resources

Per-Olov Johansson
Karl-Gustaf Löfgren

Basil Blackwell

First published 1985

Basil Blackwell Ltd
108 Cowley Road, Oxford OX4 1JF, UK

Basil Blackwell Inc.
432 Park Avenue South, Suite 1505,
New York, NY 10016, USA

British Library Cataloguing in Publication Data

Johansson, Per-Olov
The economics of forestry and natural resources.
1. Forests and forestry – Economic aspects
I. Title II. Löfgren, Karl-Gustaf
338.1′749 SD393

ISBN 0-631-14162-6

Library of Congress Cataloging in Publication Data

Johansson, Per-Olov, 1949–
The economics of forestry and natural resources.
Bibliography: p.
Includes index.
1. Forests and forestry – Economic aspects.
2. Forest products industry. 3. Natural resources – Economic aspects.
I. Löfgren, Karl-Gustaf. II. Title.
SD393.J56 1985 333.75 84-28329

ISBN 0-631-14162-6

Typeset by Unicus Graphics Ltd, Horsham, West Sussex
Printed in Great Britain by T J Press Ltd, Padstow

Contents

Preface

Scope of the Study

The object of this volume is to synthesize and integrate the existing work on forest economic theory. In a number of cases, however, this involves pushing the theory forward in new directions of treating new cases that have escaped earlier investigators, but nevertheless seem worth pursuing.

Plan of the Book

Time is of particular importance in decisions related to natural resources, and thus investment theory is necessary for the study of these decisions. The methods and criteria governing investment decisions are fairly simple and are therefore used by many. The underlying principles, although quite simple, are seldom made explicit and are therefore known by few. This is why the book opens with a chapter on investment theory.

To be understood correctly forest economics must be connected to the fundamentals of the economics of non-renewable and renewable resources. Therefore the book includes two chapters dealing exclusively with the general principles of the economics of natural resources. In chapter 2 the reader is made familiar with what is essentially the early work of Harald Hotelling on the economics of non-renewable resources, which has proved to be both correct and comprehensive. Chapter 3 introduces the economics of renewable resources, which could be apprehended as a more general version of the economics of non-renewable resources. The chapter also contains a comprehensive discussion of a well-known externality which is usually referred to as the problem of the commons.

Chapters 4–8 form a sequence dealing with pure forest economics. Chapter 4 treats the most classical problem in forest economics, i.e. the so-called rotation problem. The planning problem is to choose the age at which trees ought to be cut so as to maximize the present value of the flow of profits. In this chapter we derive the (socially) optimal rotation period, the Faustmann-Pressler-Ohlin rotation period, and investigate how it is affected by changes in prices, wages and interest rates. Other proposed cutting criteria, such as maximum sustained yield and allowable cut, are also introduced and compared with the socially optimal criterion.

Chapter 5 contains some important extensions of the problems treated in chapter 4. The effects of different kinds of taxes on the optimal rotation period and on the value of forest lands are investigated. The chapter also considers the effects on the socially optimal rotation of changes in the biotechnology and in the rate of inflation.

Chapter 6 considers a forest management problem that is more general than the Faustmann-Pressler-Ohlin problem. Explicit account is now taken of the age structure of forest stands, and timber prices are allowed to vary between periods. Within this environment, it is shown that a linear programming approach lends support to the old practice of treating the forest management problem at a stand level (as is also done in chapters 4 and 5). In the second part of the chapter, however, we abandon the assumption of constant returns to scale and introduce the more realistic assumption that the cutting technology exhibits increasing and/or decreasing returns to scale.

In chapter 7 we model the behaviour of a self-employed forest farmer who is free to allocate his working time between forestry, agriculture and industry. The chapter also contains an attempt to model the forest industry as a forest owner, which provides insights into the management of the forest industry's own forests under perfect and imperfect conditions in the roundwood market. The analysis of the forest industry also provides us with the demand function for roundwood, which is necessary for an analysis of the roundwood market.

In chapter 8 we put together the pieces from the previous chapters into a multiperiod partial equilibrium model of the roundwood market. However, perhaps the most important content of the chapter is that we allow for a monopsonistic market structure in the roundwood market. The fact that a resource base like forests is spatially rather than evenly distributed, while the forest industry is spatially concentrated, is likely to create a situation where a strong buyer of wood faces many weak sellers. This structure is modelled by using monopsony theory and elements from bargaining theory.

The only way to discriminate between different theories is to derive empirical theorems from them and then to confront these with reality. In chapter 9 a slightly modified version of the multi-period partial equilibrium model of chapter 8 is estimated using time series data for Sweden. The chapter is particularly useful in demonstrating how one converts a rather general theoretical model into an empirical model. The chapter also demonstrates that even rather simple forest economic models derived from sound economic theory may have a considerable predictive power.

Chapter 10 contains elements from welfare theory. The purpose of the chapter is to show how traditional welfare theory can be applied to answer the very important normative question of how natural resources should be priced in order that the pricing policy may be considered optimal in the Pareto sense.

At least national forest lands are often managed in a way which departs from the Faustmann-Pressler-Ohlin criterion presented in chapters 4 and 5. This is true, at least in part, because national forest lands are managed also for non-timber purposes like recreation, wilderness and wildlife. Another important reason for not adopting the Faustmann-Pressler-Ohlin rule is that the public timber supply policy is often designed to minimize the variation in employment over time for those whose employment is directly based upon national forest timber sales, and (or) to reduce temporary shortages of timber. Chapter 11 develops an intertemporal multisector model of an open economy which is used to derive cost–benefit rules to be used in such situations.

The book ends with a relatively short but important chapter on the modelling of timber supply under uncertainty. In the first part of the chapter we investigate a simplified stochastic variation of the rotation problem considered in chapter 4. We assume that value growth is stochastic – follows a Markov process – and we search for the cutting rule which maximizes the expected present value of the cut. It turns out that this rule, under fairly realistic assumptions for the distribution function of growth, is analogous to the corresponding rule under certainty.

In the second part of chapter 12, we simplify the analysis in chapter 6 by collapsing the number of periods into two – today and tomorrow. Today's prices are certain, while tomorrow's prices contain stochastic elements. A risk averse forest owner maximizes the expected utility of the cut in this setting, and the results which are obtained are in important respects different from the results obtained from a similar model under certainty. In considering the natural uncertainties involved in problems with a time horizon of

200 years and more, this final chapter hints at an important area for future research.

Acknowledgements

A great number of research environments and people have shaped the final version of this volume. The book originates from sabbatical leave in 1980 which Karl Gustaf Löfgren spent at the University of Wisconsin, Madison, learning forest economics. The stimulating research environment provided by the Poverty Institute and the Department of Economics inspired Löfgren to write a first draft of a few chapters on a book on forest economics.

Soon after Löfgren returned from Madison, Per-Olov Johansson started to do joint research work with him on forest economical matters. When in 1983 Johansson moved to the Department of Forest Economics, Swedish University of Agricultural Sciences, Umeå, the book became a joint project.

Associate Professor Håkan Persson, Department of Economics, University of Gothenburg, read the 1982 version of the manuscript and suggested a number of improvements. Assistant Professor Lance Brännman, University of Wisconsin, Milwaukee, read the 1984 version and he not only suggested a number of technical improvements but also spent a lot of time improving our Swenglish.

Graduate students at the University of Umeå and the Swedish University of Agricultural Sciences have been plagued by earlier versions of the manuscript. A non-trivial number of them have corrected a number of non-trivial errors and suggested other improvements to the text.

Different scholars have commented on parts of the manuscript. We would in particular like to mention Professor Joe Buongiorno, University of Wisconsin, Madison; Professor Björn Hägglund and Associate Professor Bo Ranneby, Swedish University of Agricultural Sciences, Umeå; Professor Ravi Kanbur, University of Essex; and Roger Sedjo, Director of Resources for the Future, Washington, DC. In addition, we acknowledge comments from scholarly people at various conferences and from anonymous referees, who all have been more or less forced to listen to or read bits and pieces of some of the material assembled in this volume.

Thanks are due also to Bill McKillop and his graduate students at Berkeley, whose reactions to some of the material in chapters 6 and 12 stimulated us to carry on.

Special thanks go to some of our own graduate students: Runar Brännlund, who very actively contributed to the creation of the empirical material presented in chapter 9; Gunnel Bångman and Anders Wiberg, who contributed to a paper that was the embryo of chapter 6; and Peter Lohmander, who has checked many of our calculations.

We are also indebted to a number of secretaries for their incredible endurance and ability to decipher hieroglyphics. The lion's share of the credit goes to Marie-Louise Harder and Marie Hammarstedt at the Department of Forest Economics, Swedish University of Agricultural Sciences; however, Eva Cederbladh, Ing-Marie Nilsson and Lisbeth Segerlund at the Department of Economics, University of Umeå, were also given a hard time by authors pressed by real or imaginary time limits. Alan Harkess, Department of Economics, University of Lund, made the final manuscript readable by scrutinizing the language.

We gratefully acknowledge research grants from the Fund for Forest Research, and the Nils and Dorthi Troedssons Fund for Forest Research, which enabled us not only to do the underlying research but also to present the material at numerous conferences.

In spite of all the help that has been provided by outsiders, there are undoubtedly errors and flaws remaining in the manuscript. As a matter of good form, we remind the reader that they are the responsibility of the authors.

1

An Introduction to the Theory of Investment

1.1 Introduction

One of the characteristic features of forest economics and the economics of natural resources is intertemporal choice, i.e. the allocation of resources for consumptive and productive purposes over time. In forest economics, the classical economic problem is to determine when a forest stand should be cut. The related problem in natural resource economics is to determine the optimal rate at which a non-renewable or renewable natural resource should be exploited.

The theory of investment deals with the principles of intertemporal choice, and the purpose of this chapter is to review some of the basic principles which are involved in generating intertemporal decision rules, thereby determining which of a number of possible investment projects should be undertaken.

We have found this return to basics necessary because forest economists sometimes discuss forestry investment decisions as if they ought to be governed by very particular rules which are dictated by the long run nature of forestry. This is exemplified by the lively discussion in Sweden and Germany at the beginning of this century[1] on the correct investment rule in forestry. This discussion went a bit astray, probably owing to the lack of an appropriate theoretical underpinning.

Another reason for this chapter is that it furnishes us with insights into situations where, under certain realistic assumptions, there may not exist an investment decision which is *objectively preferred*[2] to the remaining alternatives.

[1] A good summary of the arguments involved is found in Streyffert (1938). The discussion is often referred to as the 'profitability war'.

[2] To be defined in the appendix to this chapter.

The discussion of the theory of investment will be brief, and the reader who wants a more comprehensive analysis is referred to Hirshleifer (1970).

1.2 Investment Decisions in a Perfect Capital Market

In this section, the investment decision is considered under the assumption that the markets for intertemporal claims are perfect: there exists a unique interest rate at which the economic agent can borrow and lend any amount without (significantly) affecting the rate. An investment is meant here a series of negative and positive payments made at different points in time.

The initial question is as follows. Consider any two investments that yield different net payments over a finite period:[3]

$$a = (a_1, a_2, \ldots, a_T)$$

$$b = (b_1, b_2, \ldots, b_T) \qquad a \neq b$$

Under what conditions do criteria exist whereby any two agents would choose the same investment? An investment involves a trade-off between consumption today and consumption tomorrow. Hence, it might seem impossible to separate investment decisions from consumption decisions. However, if total income is known in advance and the individual, by borrowing and lending (saving), reallocates consumption over time within the limitations of current and future income, any individual would choose the investment yielding the largest *present value*. The reason why is simple:

> A larger present value is equivalent to a larger choice (or consumption) set, and a larger choice set will always allow a better – or at least the same – choice as the smaller choice set, over any set of periods.

More formally, assume that the individual possesses the utility function

$$U = U(c_1, \ldots, c_T) \tag{1.1}$$

where c_t is the real value of consumption at the end of period t.

[3] The payments (a_t, b_t) are assumed to yield net revenue for the same number of periods (T periods). This involves no loss of generality, as an investment that lasts only $m < T$ periods can be made to last T periods by the addition of $T-m$ zeros.

The only assumption made about the utility function is that it ranks different consumption vectors. No assumptions such as monotonicity and continuity are necessary for the argument.

Assume that all payments are made at the end of the periods. Moreover, assume that:

Assumption 1.1

(a) Future real incomes y_t $(t = 1, \ldots, T)$ are known with certainty.

(b) Net real receipts from the investments are known with certainty.

(c) The capital market is perfect in the sense that any amount can be borrowed and lent at prevailing interest rates, and that future interest rates are known with certainty.

During the first period the agent receives income y_1, consumes c_1 and invests $a_1 < 0$. The net holdings at the end of the first period are

$$y_1 - c_1 + a_1$$

During period 2 he (she) receives $y_2 + a_2$, consumes c_2 and inherits the first period's net holdings plus interest $(y_1 - c_1 + a_1)(1 + r_2)$. His net holdings at the end of period 2 can be written

$$(y_1 - c_1 + a_1)(1 + r_2) + y_2 - c_2 + a_2$$

A similar process for $t = 3, \ldots, T$ yields a growth of net holdings at the end of the planning period:

$$\begin{aligned} &(y_1 - c_1 + a_1)(1 + r_2)(1 + r_3) \ldots (1 + r_T) \\ &+ (y_2 - c_2 + a_2)(1 + r_3) \ldots (1 + r_T) \\ &+ \ldots + (y_{T-1} - c_{T-1} + a_{T-1})(1 + r_T) + y_T - c_T + a_T \end{aligned}$$

By restricting terminal net wealth to be non-negative, and dividing by $(1 + r_2)(1 + r_3) \ldots (1 + r_T)$, we obtain the intertemporal budget constraint expressed in terms of initial net wealth:

$$\begin{aligned} &(y_1 - c_1 + a_1) R_1 + (y_2 - c_2 + a_2) R_2 + \ldots \\ &+ (y_T - c_T + a_T) R_T \geqslant 0 \end{aligned} \tag{1.2a}$$

where

$$R_1 = 1, \quad R_2 = \frac{1}{1 + r_2}, \quad \ldots, \quad R_T = \frac{1}{(1 + r_2)(1 + r_3) \ldots (1 + r_T)}$$

are discount factors, and r_t is the interest rate that prevails during the tth period. The budget constraint (1.2a) says that the present value of an individual's net wealth (after consumption is deducted) is non-negative. However, if the marginal utility of real consumption is positive, the individual will never keep any real wealth beyond the planning horizon since it has no utility.

If the individual does not invest, his real consumption is constrained by

$$(y_1 - c_1) R_1 + (y_2 - c_2) R_2 + \ldots + (y_T - c_T) R_T \geqslant 0 \qquad (1.2b)$$

The relevant question is now: how can it be determined which of the two options (1.2a) or (1.2b) has the larger choice set? The present value of consumption C under the investment regime is bounded from below by zero, assuming that positive consumption is feasible, and from above by $\Sigma_{t=1}^{T} (y_t + a_t) R_t$:

$$\sum_{t=1}^{T} (y_t + a_t) R_t \geqslant C \geqslant 0 \qquad (1.3a)$$

When no investment is undertaken we have the bounds

$$\sum_{t=1}^{T} y_t R_t \geqslant C \geqslant 0 \qquad (1.3b)$$

Since the discount rates are independent of the particular consumption vector (R_t is independent of C), the option with the larger choice set has the larger upper bound. By subtracting (1.3b) from (1.3a) we see that the investment regime has the larger choice set if and only if

$$\sum_{t=1}^{T} a_t R_t > 0 \qquad (1.4)$$

i.e. if the present value of the investment is positive.

An alternative approach is to find the discount rate (the so-called internal rate of return), which would make the present value of the investment equal to zero. Define

$$K(r_1, \ldots, r_T) = \sum_{t=1}^{T} a_t R_t \qquad (1.5a)$$

and assume that the rate of interest is constant over the whole planning period. Then

$$K(r) = \sum_{t-1}^{T} a_t R_t \qquad (1.5b)$$

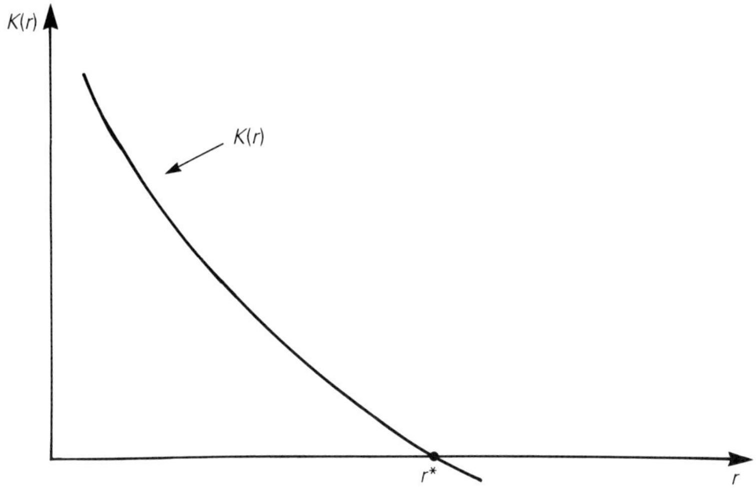

FIGURE 1.1 The internal rate of return

Furthermore, assume that:

Assumption 1.2
(a) $a_1 < 0$.

(b) $a_2, \ldots, a_T \geqslant 0$ and $\sum_{t=1}^{T} a_t > 0$.

The function $K(r)$ is a decreasing function of r. However, since $\Sigma_{t=2}^{T} a_t R_t$ is positive, but can be made arbitrarily small by choosing a sufficiently large r, there exists an $r^* > 0$ such that $K(r^*) = 0$ (see figure 1.1).

As mentioned earlier, the discount rate (r^*) that makes the present value of the net payment of the project zero is usually called the *internal rate of return*.[4] The internal rate of return, assuming a constant rate of interest,[5] can be used as a criterion for whether to invest or not to invest. If $r^* > r$, i.e. if the internal rate of return is larger than the prevailing interest rate, the present value of the invest-

[4] The rate of return is also used; see e.g. Hirshleifer (1970). In this study, however, the rate of return concept is reserved for the social discount factor.

[5] If the rate of interest varies between periods it is clearly not possible to use this simple criterion as there is no interest rate for purposes of comparison. A weighted average of $r_1, \ldots, r_T$ will not do the trick as, loosely speaking, the 'appropriate weights' will vary between individuals.

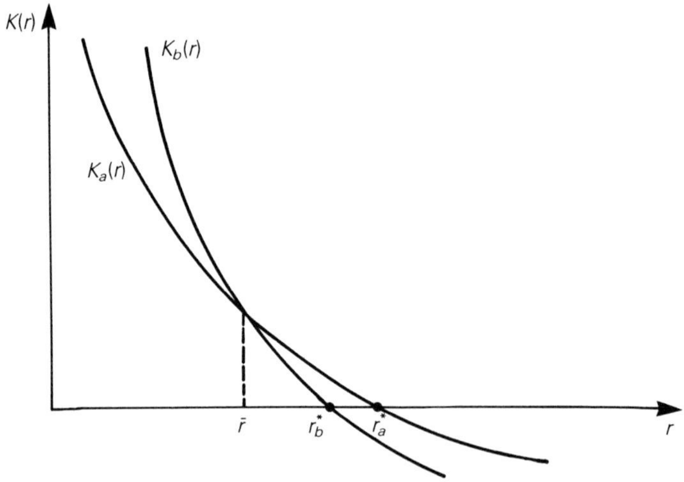

FIGURE 1.2 Why the internal rate of return does not work

ment will always be positive and the investment regime yields the larger choice set.

One can rank two investments by comparing the present value of the difference investment:

$$K_{a-b} = \sum_{t=1}^{T} (a_t - b_t) R_t = K_a - K_b \tag{1.6}$$

That is, one compares the differences in upper bounds on the discounted value of real consumption.

If the present value of the difference investment K_{a-b} is positive, investment a is preferable to investment b. It is, however, not possible to rank any two investments solely by their internal rates of return, i.e. $r_a^* > r_b^*$ does not imply that investment a is preferred to investment b. Differences in the internal rate of return do not always rank differences in choice sets correctly, because the choice sets depend on the prevailing interest rate (see figure 1.2). The internal rate of return of investment a is larger than the internal rate of return of investment b, but investment b is clearly to be preferred when the rate of interest is lower than $\bar{r}$.

In sum:

Proposition 1.1

Given assumption 1.1, any two investments can be ranked according to their present values. The investment with the larger present value yields the larger choice set. An investment with a positive present value is preferred to or equivalent to no investment.

Proposition 1.2
Given assumption 1.1 and 1.2 and a constant market interest rate, investment a is preferred or equivalent to no investment if the investment's internal rate of return is greater than or equal to the market interest rate. It is, in general, not possible to rank two investments according to their internal rates of return.

These propositions may be easily illustrated in the two period case. The budget restriction when no investment is undertaken can be written

$$y_1 - c_1 + (y_2 - c_2)(1 + r)^{-1} \geqslant 0 \tag{1.7a}$$

or

$$c_2 \leqslant y_2 + (y_1 - c_1)(1 + r)$$

The budget restriction when investment a is undertaken can be written

$$y_1 + a_1 - c_1 + (y_2 + a_2 - c_2)(1 + r)^{-1} \geqslant 0 \tag{1.7b}$$

where

$$a_1 < 0, \quad a_2 > 0$$

or

$$c_2 \leqslant y_2 + a_2 + (y_1 + a_1 - c_1)(1 + r)$$

Both budget lines are drawn in figure 1.3. Since the slopes of the two budget lines $dc_2/dc_1 = -(1 + r)$ are the same, the two choice sets can be compared for 'size' by comparing the ordinate ($c_1 = 0$) or the abscissa ($c_2 = 0$) at the origin.[6]

The differences in coordinates between points A′, B′ and A, B are

$$\begin{aligned} B' - A' &= a_1(1 + r) + a_2 = K(r)(1 + r) \qquad (c_1 = 0) \\ B - A &= a_1 + \frac{a_2}{1 + r} = K(r) \qquad (c_2 = 0) \end{aligned} \tag{1.8}$$

which are equal to the present value of the investment evaluated at the end of the second and first periods, respectively. From figure 1.3

[6] Any other point will do as well.

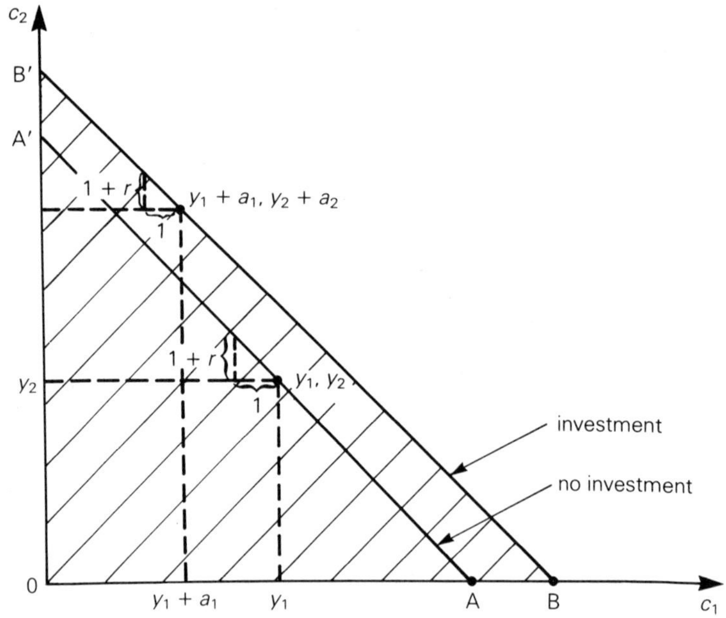

FIGURE 1.3 The choice sets in the two period case

and (1.8) we see that the choice set under investment a is larger than the choice set under no investment if and only if $K(r) > 0$. The same graphical technique can be used to compare any two investments a and b. Hence, proposition 1.1 holds in the two period case.

In the discussion so far, an investment project has been treated as a discrete choice between different income streams. This type of choice applies to the analysis of the cutting behaviour of private forest owners where the biotechnology allows for a nearly continuous spectrum of options, even after the regeneration (investment) decision has been taken. The forest owner has to decide the right time to cut a particular stand and start a new rotation. The effect of a perfect capital market in this environment is to make the cutting decision independent of the particular utility function, i.e. any utility maximizing forest owner would choose the same cutting programme.

In order to demonstrate this, assume that the biotechnology in our two period problem is described by the continuous and concave implicit function[7]

[7] This specification of the biotechnology does not enable us to treat the cutting decision as a pure rotation problem. A more elaborate presentation of the rotation problem is deferred until chapter 4.

$$F(x_1, x_2) = 0 \tag{1.9}$$

where x_t is the real value of the cut in period t.

Assume that the units have been chosen so that all consumer and timber prices equal 1. The intertemporal budget constraint of the forest owner is written

$$(x_1 - c_1)(1 + r) + x_2 - c_2 + \bar{y} = 0 \tag{1.10}$$

where $\bar{y}$ is the exogenously determined real income received in the second period.[8]

The optimization problem of the investor can now be formulated as

$$\max_{c_1, c_2} U = U(c_1, c_2) \tag{1.11}$$

subject to

$$F(x_1, x_2) = 0$$

$$(x_1 - c_1)(1 + r) + x_2 - c_2 + \bar{y} = 0$$

with Lagrangian

$$L(c_1, c_2, x_1, x_2, \mu, \lambda)$$
$$= U(c_1, c_2) + \mu F(x_1, x_2) + \lambda[(x_1 - c_1)(1 + r) + x_2 - c_2 + \bar{y}] = 0 \tag{1.12}$$

where μ and λ are Lagrange multipliers. The first order conditions for a maximum are

(a) $U_1 - \lambda(1 + r) = 0 \qquad U_t = \dfrac{\partial U}{\partial c_t}$

(b) $U_2 - \lambda = 0 \qquad F_t = \dfrac{\partial F}{\partial x_t}$

(c) $\mu F_1 + \lambda(1 + r) = 0$ (1.13)

(d) $\mu F_2 + \lambda = 0$

(e) $F(x_1, x_2) = 0$

(f) $(x_1 - c_1)(1 + r) + (x_2 - c_2) + \bar{y} = 0$

By dividing (1.13c) by (1.13d) we have

$$\frac{F_1(x_1, x_2)}{F_2(x_1, x_2)} = 1 + r \tag{1.14}$$

[8] The assumption that $\bar{y}$ is received in the second period is just a technical detail to avoid an explicit discount factor.

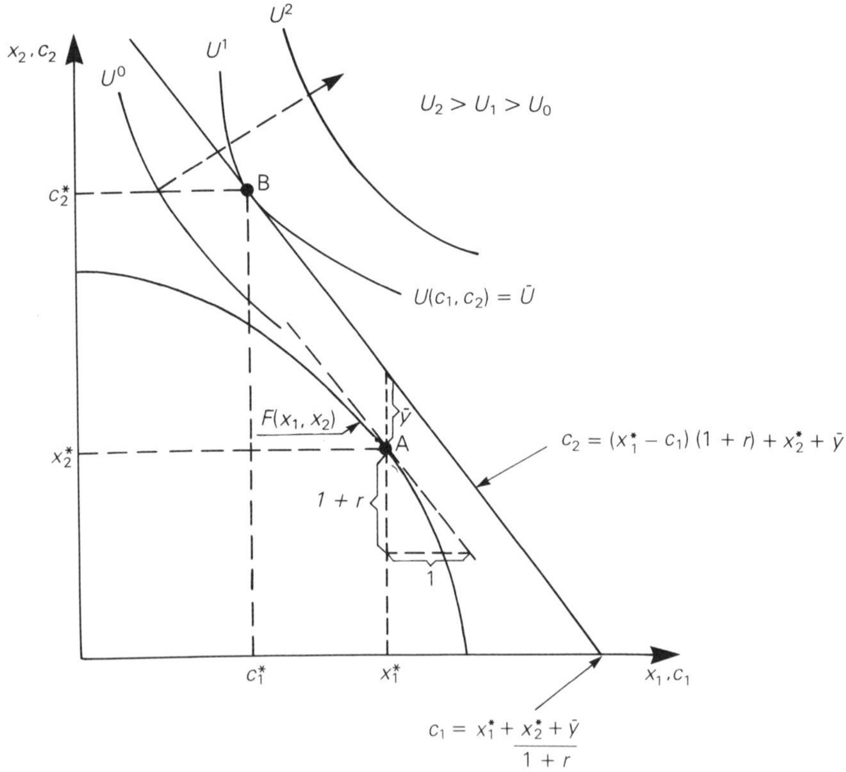

FIGURE 1.4 The separability of the cutting decision and the consumption decision $c_1 = x_1^* + [(x_2^* + \bar{y})/(1 + r)]$

Using the biotechnology in (1.13e) we can solve for x_1 and x_2 as functions of the interest rate and the prices for timber in the respective periods. Thus the optimal cutting programme is independent of the preferences of the forest owner. The solution is illustrated in figure 1.4. Point A is the *cutting decision*, determined by the equality of the slope of the transformation curve,[9] $-F_1(x_1, x_2)/F_2(x_1, x_2)$ and the slope of the budget line $-(1 + r)$. The *consumption point* B is determined by the tangency point between the budget line, generated by $x_1 = x_1^*$ and $x_2 = x_2^*$, and an indifference curve. The indifference curves U^0, U^1, U^2, i.e. the combinations of c_1 and c_2 yielding a given utility level, are negatively sloped and convex to the origin. This

[9] To see this, note that along the transformation curve $F_1(x_1, x_2)\,dx_1 + F_2(x_1, x_2)\,dx_2 = 0$ and hence $[dx_2/dx_1]_{F=0} = -F_1(x_1, x_2)/F_2(x_1, x_2)$.

shape results from an assumption of a positive marginal utility of consumption and a quasi-concave utility function.

1.3 Investment Decisions in an Imperfect Capital Market

We now allow for an imperfect capital market. A simple way to introduce capital market imperfections is to substitute the following for assumption 1.1(c):

(c′) The capital market is imperfect in the sense that the lending rate is different from the borrowing rate. Both interest rates are known with certainty, and are constant over time.

This assumption is a fairly reasonable first approximation to reality since rates paid on deposits are lower than rates paid when money is borrowed from the banking system. With an imperfect capital market, the upper bound on the present value of consumption is no longer well defined since the present value of the total income stream depends on the time pattern of consumption. The mix between borrowing or savings during one period leads to different upper bounds on the present value of consumption since borrowing rates are different from lending rates.

To treat the investment problem under an imperfect capital market in a simple, graphical way, we will assume that there are only two periods.

In figure 1.5, the real value of consumption during the first and second period is measured along the horizontal and vertical axes, respectively.

Point A in the diagram denotes the consumption possibility if saving and borrowing are not allowed (y_1 and y_2 are the incomes of the agent during the first and second period, respectively). If amount s is saved during the first period, the agent will obtain $(1 + r_1)s$ at the end of the second period, where r_1 is the lending rate. Hence, the bound on the consumption possibilities if the agent chooses to save is given by the straight line segment AB.

If the agent chooses $c_1 > y_1$ he will have to borrow the amount $(c_1 - y_1) = d$ at the interest rate r_b, and he will have to repay the amount $d(1 + r_b)$ at the end of the second period. Hence, the bound on consumption possibilities under a borrowing regime is given by the straight line AC. Since $r_b > r_1$ the budget constraint has a kink at the point A. The feasible consumption set is given by the area OBAC.

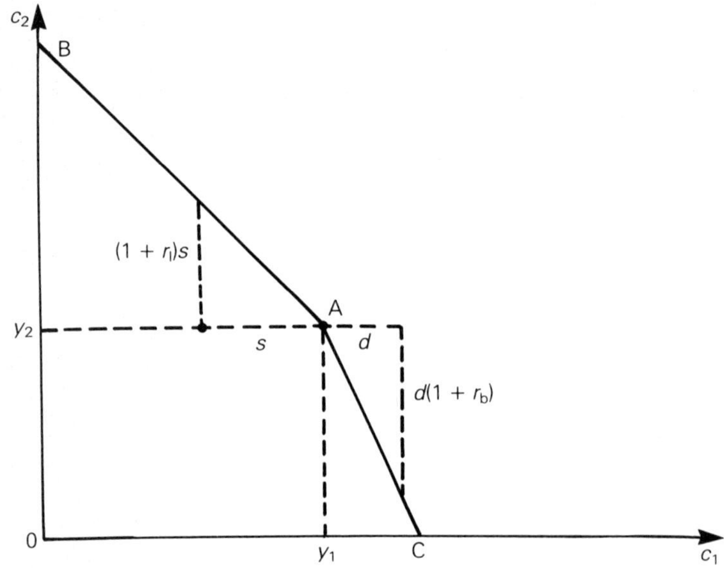

FIGURE 1.5 The choice set in an imperfect capital market

When investment is introduced, the agent makes a payment ($a_1 < 0$) at the end of the first period in the hope of receiving a payment $a_2 > 0$ in the second period. Hence the consumption possibility, if saving and borrowing are not allowed, is given by point A′ in figure 1.6.

By applying the same reasoning as in the non-investment case, it is seen that the feasible consumption set under the investment regime is given by the area OB′A′C′ in figure 1.6. Clearly a decision whether to invest or not cannot be recommended on objective grounds since there exist choices in OBAC and not in OB′A′C′, and vice versa.

When, then, is one decision better than another? If there are consumption bundles belonging to the choice set under the investment regime that do not belong to the choice set under the non-investment regime, and every consumption choice under the non-investment regime is also feasible under the investment regime, then it is obviously the case that the choice set under the investment regime is unambiguously larger, and the investment can be recommended independently of preferences. In mathematical terms one could say that an investment can be recommended if the choice set under the

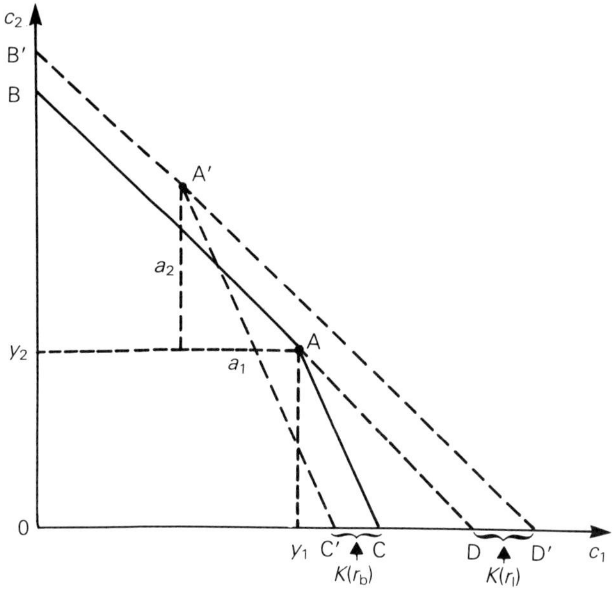

FIGURE 1.6 The feasible consumption sets under an imperfect capital market regime

non-investment regime is a proper subset of the choice set under the investment regime, i.e. if

$$S_{NI} \subset S_I \text{ (and not } S_I \subset S_{NI}) \tag{1.15a}$$

and cannot be recommended if

$$S_I \subset S_{NI} \text{ (and not } S_{NI} \subset S_I) \tag{1.15b}$$

where S_{NI} and S_I are the feasible choice sets under the non-investment and the investment regime, respectively. If neither (1.15a) nor (1.15b) holds, no decision can be recommended without knowledge of the agent's preference (utility function).

Relationships (1.15a) or (1.15b) are, however, very general and the conclusions are in a sense trivial. What one needs is a criterion to determine whether (1.15a) or (1.15b) holds.

In figure 1.7 the decision to invest is clearly preferable to the decision not to invest, i.e. ($S_{NI} \subset S_I$). The difference between the coordinates of the points C′ and C is

$$K(r_b) = a_1 + a_2(1 + r_b)^{-1} > 0 \tag{1.16a}$$

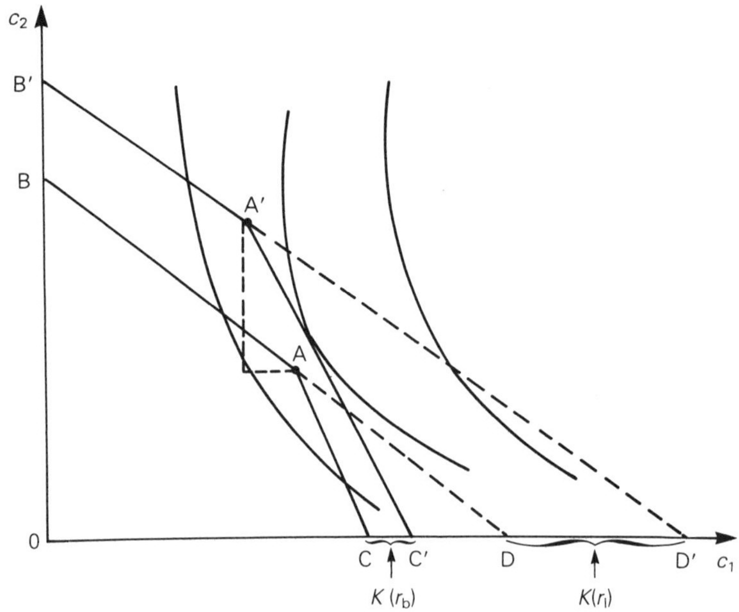

FIGURE 1.7 A profitable investment under an imperfect capital market

which is equal to the present value of the investment evaluated at the borrowing interest rate. The difference between the coordinates of the points D′ and D is

$$K(r_l) = a_1 + a_2(1 + r_l)^{-1} > 0 \tag{1.16b}$$

which is the present value of the investment evaluated at the lending interest rate. Clearly, the conditions $K(r_b) > 0$, $K(r_l) > 0$ are necessary and sufficient for the investment to be preferred to the non-investment alternative, i.e. the present values of the investment calculated in the two different possible ways must be positive.

The same criterion will also work where two investment regimes are compared. Investment a is preferred to investment b, if

$$K_{a-b}(r_b),\ K_{a-b}(r_l) > 0.$$

This ranking system holds only for the two period case. What happens if the investment problem is considered in a T period setting? Depending upon how the periodization is made, under the T period case there are 2^{T-1} or 2^T different present values to be calculated. If the end of the first period is the first point of time there are 2^{T-1}

calculations. Intuitively, an investment is preferred to no investment if all these present values are positive. This, in fact, is true and we now *sketch* the proof of this proposition (first proved by Puu, 1964; see also Puu, 1967).

The present value of the investment measures the difference in the upper bounds of the present value of the consumption bundles when the investment is compared with no investment. Thus, the fact that all 2^{T-1} possible present values of the investment are non-negative means that the least upper bound is larger under the investment regime independently of the consumption pattern over time, i.e. independently of the relevant interest rate pattern. This implies that any feasible consumption pattern under the non-investment regime is also feasible under the investment regime, and sufficiency follows.

On the other hand, if the present value of the investment is negative for at least one interest rate pattern, consumption patterns exist which are feasible under the non-investment regime and non-feasible under the investment regime. Hence, the investment cannot be recommended without knowledge of the preferences for consumption patterns over time – the necessity argument.

A similar argument can be applied to the case when two investment regimes are compared since the case of no investment is a special case with a payment structure $(0, 0, \ldots, 0)$. The results are summarized in the following proposition:

Proposition 1.3 (Puu, 1964)
Given assumption 1.1 with imperfect capital markets (c′), any two investments can be objectively ranked according to the present value criterion if and only if the 2^{T-1} possible different present values of the difference investment $a-b$ have the same sign.

Proposition 1.3, which is proved in the appendix to this chapter, may seem very restrictive and difficult to apply. However, if a system exists that selects the interest rate combination which minimizes the present value of the income stream, it is sufficient to check whether the minimum is positive or negative. Clearly, if the minimum present value is positive, the investment can be recommended. It is not possible to recommend the investment, however, if the minimum is negative. To reject the investment one has to show in addition that the maximum of the present value is negative.

A maximum or minimum of the present value is easily found if the income stream consists of one negative payment at the end of period zero and non-negative payments at the end of all the other periods. The minimum is obtained by using the highest of the prevailing interest rates as the rate of discount. The maximum emerges if the smallest of the prevailing interest rates is used. An algorithm that does this for any payment pattern is introduced in the appendix to this chapter.

Capital markets, of course, may be imperfect in respects other than differences between borrowing and lending rates. There can, for example, be restrictions on the amount that it is possible to borrow or lend.[10] An analysis of the investment decision under these circumstances will give results which imply that it might be possible to recommend a decision in the objective sense depending on whether (1.15a) or (1.15b) is fulfilled. The check on whether these conditions are fulfilled might, however, become very difficult.

To see that the cutting decision is not independent of the consumption decision in an imperfect capital market, assume that the interest rate is an increasing (decreasing) function of the amount borrowed (saved) i.e. $r=r(x_1-c_1)$ and $\partial r/\partial x_1<0$. By an appropriate reformulation of the perfect capital market problem, the reader can easily convince himself that the optimal cutting programme is not independent of the consumption programme. The first order conditions of the modified maximization problem will have the following form:

$$\text{(a)}\quad U_1-\lambda\left[(1+r)-(x_1-c_1)\frac{\partial r}{\partial c_1}\right]=0$$

$$\text{(b)}\quad U_2-\lambda=0$$

$$\text{(c)}\quad \mu F_1+\lambda\left[(1+r)+(x_1-c_1)\frac{\partial r}{\partial x_1}\right]=0 \qquad (1.13')$$

$$\text{(d)}\quad \mu F_2+\lambda=0$$

$$\text{(e)}\quad F(x_1,x_2)=0$$

$$\text{(f)}\quad (x_1-c_1)(1+r)+(x_2-c_2)+\bar{y}=0$$

Equations (1.13′c–e) alone cannot be used to solve for x_1, x_2, since r and $\partial r/\partial x_1$ depend on c_1. In other words, (1.13′c–e) form a system of three equations in four variables.

[10] For investment decisions under a borrowing restriction, a recent paper by Cantor and Lippman (1983) provides some new insights. They show that a rate of return criterion maximizes terminal wealth, i.e. wealth at planning horizon.

1.4 Defining a Continuous Discount Rate

The present value computations making use of the discount factors $R_1, R_2, \ldots, R_T$ are quite easy since there are tables that supply the values of R_T for most interest rates and maturity dates. In theoretical work, however, there are many advantages in using a continuous approach – in particular, a more concise notation and a simplification of the calculus involved. Most continuous calculuses, including this one, rely heavily on the exponential function in expressing present values and the discount factor.

The relationship between the continuous discount factor e^{-rt} and the discrete discount factor $[(1+i)^t]^{-1} = R_t$ may be grasped from the following reasoning. Consider a period of one year, which is divided into a very large number n of elementary periods. If r is the continuous interest rate, then r/n is the interest earned on one dollar during the very short period n^{-1}. At the end of the year this dollar will be worth

$$v = \left(1 + \frac{r}{n}\right)^n \tag{1.17}$$

If the periodization is made infinitely minute, v converges to e^r, i.e.[11]

$$\lim_{n \to \infty} v = e^r \tag{1.17a}$$

where e is the base of natural logarithms.

The continuous annual interest rate r corresponding to the discrete discount rate i is thus defined by the formula

$$e^r = 1 + i \tag{1.18}$$

Hence, using the continuous discount factor, the present value of one dollar at time t equals e^{-rt}. If $a(t)$ is a continuous return function rather than a discrete sequence of returns $(a_1, \ldots, a_T)$, the expression

$$PV = \sum_{j=1}^{T} \frac{a_j}{(1+i)^j} \tag{1.19a}$$

is equal to

$$PV = \int_0^T a(t)\, e^{-rt}\, dt \tag{1.19b}$$

[11] The reader familiar with calculus will recognize that $\lim_{n \to \infty} [1 + (1/n)]^n = e$.

If r varies over time, rt in the exponential is replaced by

$$R(t) = \int_0^t r(x)\,\mathrm{d}x \tag{1.20}$$

and the present value of the income stream between 0 and T is

$$\int_0^T a(t)\,\mathrm{e}^{-R(t)}\,\mathrm{d}t \tag{1.21}$$

In the following chapters, frequent use will be made of the continuous approach, and the time horizon will often be infinite, i.e. $T \to \infty$.

Appendix: The Proof of Puu's Theorem and the Derivation of an Algorithm

The purpose of this appendix is to prove Puu's theorem on investment decisions in an imperfect capital market, and to prove a lemma which considerably simplifies the application of Puu's theorem to everyday investment calculations. Odhnoff's (1965) proof of Puu's theorem is modified to provide us with the appropriate lemma. We start with a few definitions. Let

$$z = (z_1, \ldots, z_T) \tag{1.22}$$

be a T tuple of real numbers. This may be interpreted as the real income stream generated by an investment z. In addition, let

$$\Omega = \{r = (r_1, \ldots, r_{T-1}):\quad r_j = r_1 \text{ or } r_j = r_b\} \tag{1.23}$$

i.e. Ω is the set of $T-1$ tuples of interest rates, where the interest rate during a particular period is equal to the saving rate (r_1) or the borrowing rate (r_b), and $r_b > r_1 > -1$. Define also

$$K_r(z_T) = z_T \tag{1.24a}$$

and, recursively,

$$K_r(z_j, \ldots, z_T) = z_j + K_r(z_{j+1}, \ldots, z_T)\,(1 + r_j)^{-1} \tag{1.24b}$$

In addition, define

$$K(z) = K_r(z_1, \ldots, z_T) \tag{1.24c}$$

Now introduce the following assignment rule:

$$r_j(z) = \begin{cases} r_b & \text{if } K_{r(z)}(z_{j+1}, \ldots, z_T) > 0 \\ r_1 & \text{if } K_{r(z)}(z_{j+1}, \ldots, z_T) \leqslant 0 \end{cases} \tag{1.25}$$

Equation (1.25) picks a certain r tuple in Ω. The assignment rule is interpreted as follows. Start with payment z_T. If $z_T > 0$, discount z_T, using the highest interest rate, to the end of period $T-1$. If $z_T \leqslant 0$, use the lowest interest rate as the discount rate.

Check the sign of the expression

$$K_r(z_{T-1}, z_T) = z_{T-1} + z_T[1 + r_T(z)]^{-1}$$

and choose the discount rate in accordance with (1.25) etc. In this way, positive present values discounted one period backwards will be made as small as possible by the use of the highest discount rate, and negative present values will be made as small as possible by the use of the lowest discount rate.

Finally, define the boundary set

$$B = \{z\colon K_{r(z)} \geqslant 0\} \tag{1.26}$$

This set contains all payment streams z which generate non-negative present values when the discount rate is chosen in accordance with assignment rule (1.25). Given the way the r tuple $r(z)$ is chosen, one can prove the following lemma:

Lemma 1.1

$$K_{r(z)} = \min_{r \in \Omega} K_r(z)$$

Proof

For $r \in \Omega$ one has that:

(a) $K_r(z_T) \geqslant K_{r(z)}(z_T)$ since both sides equal z_T.

(b) Assume

$$K_r(z_{j+1}, \ldots, z_T) \geqslant K_{r(z)}(z_{j+1}, \ldots, z_T)$$

for $j + 1 = k + 1$.

(c) Then

$$K_r(z_k, \ldots, z_T) - K_{r(z)}(z_k, \ldots, z_T)$$
$$= (1 + r_k)^{-1} K_r(z_{k+1}, \ldots, z_T) - [1 + r_k(z)]^{-1} K_{r(z)}(z_{k+1}, \ldots, z_T)$$
$$\geqslant K_{r(z)}(z_{k+1}, \ldots, z_T) \frac{[r_k(z) - r_k]}{(1 + r_k)[1 + r_k(z)]} \geqslant 0$$

The last term is non-negative by the definition of $r_k(z)$, and the fact that $r_b > r_l > -1$. The lemma now follows by induction.

One obvious interpretation of the lemma is that the assignment rule $r(z)$ selects the smallest present value generated by the real income stream z among the 2^{T-1} possible present values. Given Puu's theorem, formulated and proved later, we get a simple algorithm to determine whether an investment is 'objectively' preferable to another investment.

The boundary set is characterized by the following theorem:

Theorem 1.1 (Odhnoff)
The boundary is determined by the formula

$$B = \bigcap_{r \in \Omega} \{z: K_r(z) \geqslant 0\}$$

Proof
The inclusion

$$\bigcap_{r \in \Omega} \{z: K_r(z) \geqslant 0\} \subset B$$

is trivial.

The inclusion

$$B \subset \bigcap_{r \in \Omega} \{z: K_r(z) \geqslant 0\}$$

follows from the lemma, because if $z \in B$ the lemma implies that $K_r(z) \geqslant K_{r(z)} \geqslant 0$ for every $r \in \Omega$. This proves the theorem.

We now introduce a formal definition of 'objectively' preferred.

Definition of objectively preferred
An investment described by the T tuple $x^1 = (x_1^1, \ldots, x_T^1)$ of real payments is said to be objectively preferred to another investment $x^2 = (x_2^2, \ldots, x_T^2)$ if there exists a T tuple $\epsilon = (\epsilon_1, \ldots, \epsilon_T)$ with $\epsilon_t \geqslant 0$, $\epsilon_t > 0$ for at least one t, such that

$$(x^2 - u) \in B \text{ implies } (x^1 - u - \epsilon) \in B \text{ for every } u \in R^T$$

A possible interpretation of this definition is the following. If $u = c - y$, then u can be interpreted as the net consumption vector, where $c = (c_1, c_2, \ldots, c_T)$ is the real consumption vector and $y = (y_1, y_2, \ldots, y_T)$ is the real income vector; y is exogenously given. Hence, $x^2 - u \in B$ means that the consumption vector c is feasible, given real income y, under the investment x^2, even if the present value is calculated in the most unfavourable way. Feasible means that c fulfils the budget restriction $K_{r(z)}(x^2 + y - c) \geqslant 0$. An investment x^1 is objectively preferred to an investment x^2 if, whenever c is feasible under x^2, a slightly higher consumption $c+\epsilon$ is feasible under investment x^1. In sum:

Proposition 1.4
An investment x^1 is objectively preferred to another investment x^2 if $K_r(x^2-u)\geqslant 0$ implies $K_r(x^1-u)>0$ for all $r\in\Omega$, $u\in R^T$.

Proof
$x^2-u\in B$ implies $(x^1-u-\epsilon)\in B$, every $u\in R^T$, is equivalent to $K_r(x^2-u)\geqslant 0$ implies $K_r(x^1-u-\epsilon)\geqslant 0$, all $r\in\Omega$, $u\in R^T$, by theorem 1.1. As $K(\epsilon)>0$ and $K(x^1-u-\epsilon)=K(x^1-u)-K(\epsilon)$, an investment x^1 is objectively preferred if $K(x^2-u)\geqslant 0$ implies $K(x^1-u)>0$.

We are now ready to state and prove the main theorem:

Theorem 1.2 (Puu, 1964)
The condition $K_r(x^1)>K_r(x^2)$, i.e. $K_r(x^1-x^2)>0$, for every $r\in\Omega$ is necessary and sufficient for investment x^1 to be objectively preferred to investment x^2.

Proof
Sufficiency If $K_r(x^1)>K_r(x^2)$ for all $r\in\Omega$ and $K_r(x^2-u)\geqslant 0$ for all $r\in\Omega$, and $u\in R^T$, then $K_1(x^1-u)=K_r(x^1)-K_r(u)>K_r(x^2)-K_r(u)=K_r(x^2-u)$ all $r\in\Omega$, and $u\in R^T$ which, considering proposition 1.4, proves sufficiency.

Necessity If x^1 is objectively preferred to x^2, by proposition 1.4 $K_r(x^2-u)\geqslant 0$ implies $K_r(x^1-u)>0$ all $r\in\Omega$, $u\in R^T$. Choose $u=x^2$, and one immediately has $K_r(x^1-x^2)>0$ for all $r\in\Omega$.

Theorem 1.2 also contains the investment criterion for the perfect capital market as a special case ($r_b=r_1$).

2

The Theory of Non-renewable Resources

2.1 Introduction

This chapter contains a fairly short review of the economic theory of non-renewable (exhaustible) resources.[1] A *non-renewable resource* is a resource available in a physically limited quantity, which does not grow or cannot be augmented by a 'technical' production process. Moreover, if used as an input in a production process, the intertemporal sum of services provided by a given stock of an exhaustible resource is finite.

The scarcity of natural resources in the non-renewable sense was recognized and analysed by the classical English economists, Thomas Malthus and David Ricardo, at the beginning of the nineteenth century. They were concerned with the problems which would be created by a limited amount of cultivable land.[2] Physically limited stocks of mineral deposits and land would – according to both Malthus and Ricardo – create severe problems for mankind.

Subsequent economic development has shown that the classical economists were wrong. New discoveries, technological progress, economies of use, substitution possibilities, and the recycling of used material have made these physical resource constraints inoperative over a very long period. There are, in fact, studies which indicate that relative prices (a scarcity index) of minerals and other non-renewable resources have declined during the twentieth century (Barnett and Morse, 1963). We will not dwell on the conditions

[1] For a more comprehensive analysis the reader is referred to Dasgupta and Heal (1979), in particular chapters 6, 7, 10, 11 and 12. See also Fisher (1981).

[2] Note that land is not a non-renewable resource in the sense of the previous paragraph as land can at least in principle provide an unbounded intertemporal sum of services. On the other hand one will never the able to recovery the entire amount of an ore through recycling. There is always a leakage.

under which the relative price of physically limited resource is expected to increase or decrease over time (this will be dealt with in chapter 10). However, it is seen from the almost classical supply and demand diagram that a decreased commodity supply *ceteris paribus* leads to a higher price. The fact that relative prices of non-renewable resources have not increased indicates that commodity supplies have, in a sense, been augmented (for example through technical progress and new discoveries) relative to the demand side.

The recent increase in crude oil prices might be interpreted as an indication of increased oil scarcity. This interpretation is probably not quite correct. Increased oil prices in real terms are more likely caused by the potential for OPEC's use of its dominant power as a producer rather than as a consequence of increased resource scarcity, although we may eventually have to deal with the 'pure' scarcity problem.

The reader should also note that crude oil is not a purely non-renewable resource. The stock of oil grows, but so slowly that growth is of no practical relevance. This is, to a more limited extent, also true for a few species of fish, e.g. the blue whale (see e.g. Hannesson, 1978; Dasgupta and Heal, 1979, chapters 3 and 5). The rate of growth of blue whale stock has been insufficient to prevent the depletion of the animal, by fishery firms, almost to extinction. A profit maximizing fishery does not, as a rule, assign any value to blue whales as living creatures in a natural environment. If society assigns such a value then the private rate of exploitation will be higher than the socially optimal rate, and some kind of regulation may be required. The reasons for regulation are twofold. First, slow growth of the stock could mean that extinction is the overall optimum from a pure economic standpoint. The second motive for regulating the catch is the so-called problem of the commons. If fishing vessels have free access to a fishing ground then, under competitive conditions, overfishing occurs in the sense that the catch and the number of vessels are in excess of the socially optimal levels.[3] An extra vessel increases the catch of the firm to which it belongs, and also inflicts external diseconomies on other firms' catches by raising their costs of obtaining a given site catch. The harm done by each additional vessel is small, but the sum of many small quantities may be quite large. We will return to this problem in chapter 3, where renewable resources are analysed. It should, however, be pointed out that the problem of the commons is in no way a particular problem of renewable resources. The same problem is

[3] This problem was recognized by Gordon (1954).

present, for example, if an oil well is simultaneously exploited by two competing companies.[4]

Our review of non-renewable resource analysis is necessary in this study of forest (renewable) economics since the economic analysis of non-renewable resources reveals principles that are relevant to the economic analysis of renewable resources. In chapter 3, for example, it will be shown that the economic theory of non-renewable resources is just a special case of the theory of renewable resources.

The purpose of this chapter is to derive a couple of fundamental results regarding the properties of the economically efficient natural resource extraction path under competitive conditions, and to state conditions under which the supply of the non-renewable resource will be exhausted in finite time. It will also be shown, within a simplified model, that the economically efficient extraction path under competitive conditions depletes the resources at a socially optimal rate. A few words will also be said about how the extraction path under monopolistic conditions diverges from the socially optimal path.

2.2 A Formal Model

We start by introducing a formal model of the maximization problem of a firm extracting a non-renewable resource. This problem will first be solved intuitively, and in the latter half of the chapter more formally. In the more formal solution some use is made of optimal control theory; this is a fairly advanced mathematical technique, and readers not familiar with this technique may skim these passages. They should be read, however, since they contain optimality conditions that are earlier derived more intuitively.

Let the initial stock of the non-renewable resource be denoted $\bar{x}$, and define $x(t)$ as the remaining stock of the resource at time t. It obviously holds that

$$\bar{x} = x(0) \tag{2.1}$$

and

$$\dot{x}(t) = -h(t) \qquad [h(t) \geqslant 0] \tag{2.2}$$

where $h(t)$ is the quantity extracted at time t, and $\dot{x}(t) = \mathrm{d}x(t)/\mathrm{d}t$ is the rate of increase of the resource stock.

[4] The analysis on how to correct the market failure in this case, however, differs a little from the remedy to the overfishing problem. See Dasgupta and Heal (1979), chapter 12.

Equation (2.1) states that the stock at time zero equals the initial amount of the resource. Equation (2.2) describes how the resource stock changes over time through the extraction policy of the firm.

At any given time, it must also hold that the resource stock cannot be negative. Given that $h(t) \geqslant 0$, this condition is equivalent to

$$\int_0^\infty h(t)\,\mathrm{d}t \leqslant \bar{x} \tag{2.3}$$

Equation (2.3) says that resource extractions, summed over the planning period, cannot exceed the initial amount of the resource. The planning horizon is here set equal to infinity but, as will be seen, this does not mean that the resource is not exhausted in finite time.

Note that if we allow recycling, this does not change the underlying structure of the problem as long as there is a leakage $\alpha(0 < \alpha < 1)$. Every time h units are extracted, only $(1-\alpha)h$ units are recovered. Hence, the right hand side of (2.3) is augmented by a factor α^{-1} although the multiplier is finite.

Let the resource be owned by a firm whose objective is to maximize present value of the income stream from extraction. (The theoretical justification of this objective was discussed at some length in chapter 1.) Moreover, let the net price (net of extraction costs) of one resource unit at time t be given by the continuous function

$$p = p(t) \tag{2.4}$$

The net profit (and hence prices and extraction costs) of a unit of the extracted resource is independent of the amount exploited. This interpretation is a little unrealistic, but may be relaxed without considerable difficulty. The following intuitive reasoning is, however, much simplified by this assumption.

The present value of the income stream from extraction can be written[5]

$$PV = \int_0^\infty p(t)\,h(t)\,\mathrm{e}^{-rt}\,\mathrm{d}t \tag{2.5}$$

where r is the interest rate ($r > 0$).

The interest rate should be viewed as the riskless rate of return of a dollar invested in the ordinary sector of the economy. The rate

[5] The choice of the infinite planning horizon will be further discussed in chapter 6.

of return solves the equation

$$1 = \int_0^\infty a \, \mathrm{e}^{-rs} \, \mathrm{d}s \tag{2.6}$$

where a is a constant stream of net revenues[6] which is generated by a dollar invested in the ordinary sector. Solving for r,

$$1 = \left(-\frac{a}{r} \, \mathrm{e}^{-rs}\right)_0^\infty = \frac{a}{r} \tag{2.6a}$$

Hence

$$r = a$$

Since r by definition is the rate of return in the ordinary sector, it is the rate of discount (the opportunity cost) used to discount future revenues.

Equation (2.6) avoids many delicate problems on how the rate of return is determined in a more complicated model. The rate of discount in the real world depends in a fairly complex manner on factors such as interactions between tastes (time preference), technology, taxes and endowments.[7]

The firm's management problem, within the model defined by (2.1)–(2.5), is to choose $h(t)$, an extraction policy over time, subject to restrictions (2.1)–(2.4), such that the goal function (2.5) is maximized. Formally,

$$\max_{h(t)} \int_0^\infty p(t) \, h(t) \, \mathrm{e}^{-rt} \, \mathrm{d}t \tag{2.7}$$

subject to

$$\bar{x} = x(0)$$

$$\dot{x} = -h(t)$$

$$\int_0^\infty h(t) \, \mathrm{d}t \leqslant \bar{x}$$

$$h(t) \geqslant 0$$

[6] In an economy where the proceeds from an investment are taxed at rate s_0, the internal rate of return should be corrected for taxes by substituting $a(1 - s_0)$ for a.

[7] The interested reader is referred to Solow (1963), Koopmans (1965), and Arrow and Kurz (1970). See also Diamond and Mirrlees (1971), Drèze (1974), Sandmo and Drèze (1971).

2.3 Some Simple and some Intuitive Solutions

The absence of an upper bound on the rate of extraction $h(t)$ means that the firm is assumed to possess an unlimited extraction capacity. Under this condition, the exogenously given path of prices $p = p(t)$ reduces the optimization problem (2.7) to a pure inventory problem, i.e. the determination of the optimal time to dispose of the stock $\bar{x}$.

2.3.1 *Some Illustrations*

A special case is obtained if $p(t) = \bar{p}$ = constant. The present value of the stock is then the exponentially decreasing curve drawn in figure 2.1. In this case, it is optimal to dispose of the resource at time zero since no future price increases will be realized.

Note that the fact that net profit is independent of the rate of extraction means that the most profitable time to dispose of one resource unit coincides with the time when it is most profitable to dispose of the entire stock. In other words, with $p = p(t)$, the present value maximization problem can be written

$$\max_t PV = p(t)\,\bar{x}\,e^{-rt} \tag{2.8}$$

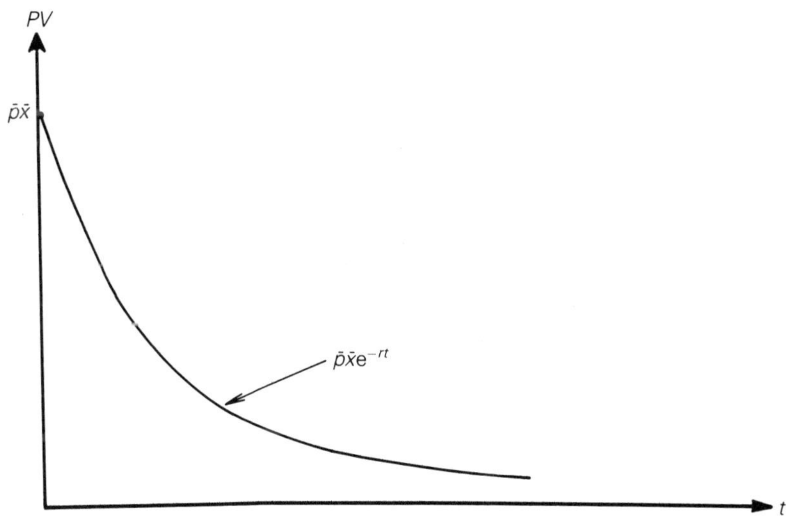

FIGURE 2.1 The present value of the resource when $p(t) = \bar{p}$ = constant

If $p(t)$ is a differentiable function of time, the first order condition for an interior maximum is

$$0 = \frac{\mathrm{d}PV}{\mathrm{d}t} = rp(t)\,\bar{x}\,\mathrm{e}^{-rt} - \dot{p}(t)\,\bar{x}\,\mathrm{e}^{-rt} \tag{2.9}$$

A few simple manipulations, assuming $p(t)\bar{x} \neq 0$, yield the following extraction rule:

Proposition 2.1
In an interior solution[8] ($T^* > 0$), the optimal extraction date T^* for the resource model specified by (2.9) is determined by the equation

$$\frac{\dot{p}(T^*)}{p(T^*)} = r$$

The economics behind this rule is fairly intuitive. If the rate of price change is greater than the interest rate, it pays to hold on to the resource, since the interest income earned from sale proceeds is less than the relative value growth of the stock. Using similar reasoning, it is unprofitable to keep the stock when $\dot{p}/p < r$.

A different interpretation of the solution can be obtained by solving a related problem with the same solution in t:

$$\max_t \log PV \tag{2.10a}$$

subject to the restriction

$$\log PV + rt \leqslant \log p(t) + \log \bar{x} \tag{2.10b}$$

That is, the natural logarithm of present value is maximized, subject to the restriction that the logarithm of present value at time t is less than or equal to the logarithm of the stock value at time t.

Figure 2.2 shows that the optimal time to dispose of the resource is determined by the tangency between the right and left hand sides of (2.10b). Any alternative investment policy, such as the one implied by $t = t_1$, yields a lower present value. Taking antilogarithms of both members of (2.10b) at $t = T^*$ yields

$$PV^*\,\mathrm{e}^{rT^*} = p(T^*)\,\bar{x} \tag{2.11}$$

or

$$PV^* = p(T^*)\,\bar{x}\,\mathrm{e}^{-rT^*} \tag{2.11a}$$

[8] Note that $\dot{p}(t) = 0$ at all t implies that (2.9) has no solution. The solution T^* is a global optimum provided that the goal function is strictly concave.

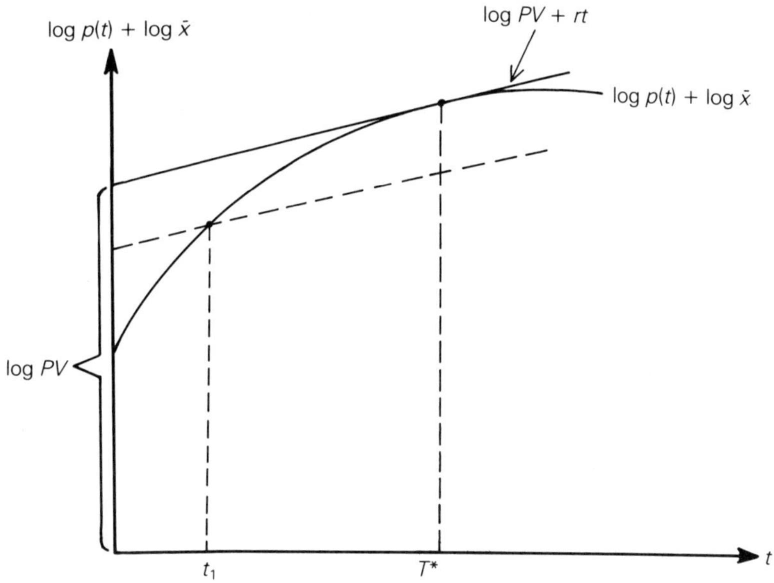

FIGURE 2.2 The determination of the optimal inventory policy

A similar management problem arises if p is constant while the stock is growing over time, i.e. $\dot{x}(t) > 0$. The optimal management problem is then formulated as

$$\max_t PV = \bar{p}x(t)\, e^{-rt} \tag{2.12}$$

with the first order condition for maximum

$$\frac{\dot{x}(T^{**})}{x(T^{**})} = r \tag{2.13}$$

Equation (2.13) says that the stock should be liquidated when its relative growth coincides with the interest rate. A possible interpretation of $x(t)$ is a forest stand (or a tree) which grows at the rate of $\dot{x}(t)$. The solution T^{**} can then be interpreted as an optimal rotation period. As will be seen in chapter 4, this solution of the rotation problem is not quite correct. However, it is readily apparent that there is a close connection between the management of renewable and non-renewable resources.

This connection is made even clearer by denoting

$$\bar{p}x(t)\, e^{-rt} = H(t)\, e^{-rt}$$

where p is constant while the stock changes over time, and

$$p(t)\,\bar{x}\,\mathrm{e}^{-rt} = G(t)\,\mathrm{e}^{-rt}$$

where p is flexible while the stock is constant.

The optimum now requires

$$\begin{aligned} \frac{\dot{H}(T^{**})}{H(T^{**})} &= r \\ \frac{\dot{G}(T^{*})}{G(T^{*})} &= r \end{aligned} \qquad (2.14)$$

Both formulas have the following common interpretation. The resource should be disposed of when its rate of increase in value coincides with the interest rate. An increase in value equal to the interest rate takes place when either the volume of the resource increases or its price increases. Then from (2.14), we see that the optimality conditions look the same.

Assume that the non-renewable resource extraction problem is solved and the optimal extraction path implies that the resource is exhaused at time T. Let

$$\lambda = p(T)\,\mathrm{e}^{-rT} \qquad (2.15)$$

be the present value of a unit of the resource at the end of the planning horizon. Assume further that the present value of a unit of the resource at an earlier time $t < T$ is less than λ, i.e. $p(t)\,\mathrm{e}^{-rt} < \lambda$. Is extraction optimal at time t? Intuitively, if the resource is extracted during an interval (t_0, t_1), and the rate of extraction may be varied, then the present value of the resource is constant over the interval. Otherwise it would be possible to increase the resource's present value by reallocating the extraction amounts from periods with low unit present values to periods with high unit present values. The only possible way to obtain a constant unit present value over an interval is if $p(t)$ fulfils the condition

$$p(t)\,\mathrm{e}^{-rt} = k \qquad t \in (t_0, t_1) \qquad (2.16)$$

i.e. $p(t) = k\,\mathrm{e}^{rt}$, which means that the net price increases at a rate equal to the interest rate. Thus, we arrive at the famous Hotelling's rule:

Proposition 2.2 (Hotelling's rule)

If a resource stock $\bar{x}$ can be extracted by the competitive firm described by (2.1)–(2.5) at a rate $0 \leqslant h(t) < \infty$ and extraction in fact takes place over an interval (t_0, t_1), then it must hold that $\dot{p}(t)/p(t) = r$ for $t \in (t_0, t_1)$.

Proposition 2.2 was first derived by Harald Hotelling in his famous (1931) paper and is proved by differentiating (2.16) with respect to t.

2.3.2 *Limited Extraction Capacity*

We now complicate the non-renewable resource extraction problem by assuming that the firm's extraction capacity is limited from above and below. More specifically,

$$0 \leqslant h(t) \leqslant h_{max} \tag{2.17}$$

When capacity is limited from above, it may no longer be possible to equalize unit present values between periods, owing to capacity limitations. On the other hand, it is not optimal to extract at $t<T$ if $p(t)\,e^{-rt}<\lambda$. Since $p(t)$ is a continuous function, it is possible to save the resource until $T+\delta$ and obtain a present unit value larger than $p(t)\,e^{-rt}$, provided that δ is made small enough.[9] This contradicts the assumption that T was optimally chosen in the first place, and proves by contradiction that it is not optimal to extract if $p(t)\,e^{-rt}<\lambda$.

The extraction rule which follows from this reasoning is: extract when $p(t)\,e^{-rt} \geqslant \lambda = p(T)\,e^{-rT}$ and save the resource when $p(t)\,e^{-rt} < \lambda$, i.e. extract when the present value of a unit of the resource exceeds the present value of the resource at the endogenous time horizon, and save the resource if the opposite holds.

The next question to ask is whether or not one should extract at full capacity utilization during intervals when extraction takes place. The answer is clearly yes, because the net price or unit profit is independent of the rate of capacity utilization. Hence, if it is optimal to extract one unit, it is also optimal to extract as many units as possible.

Our somewhat intuitive reasoning is condensed in the following proposition:

Proposition 2.3 (Bang-bang control)[10]
If a resource stock $\bar{x}$ can be extracted by a competitive firm, subject to (2.1)–(2.5) at a rate $0 \leqslant h(t) \leqslant h_{max}$, then the

[9] Note that capacity limitations are assumed to be a hindrance for the extraction of the resource at T. Moreover, if $f(t) = p(t)\,e^{-rt}$ is continuous at T, it holds by definition that for all $\epsilon > 0$ there exists a $\delta > 0$ such that $f(t) - f(T) < \epsilon$ for all t such that $t - T < \delta$.

[10] The terminology is borrowed from optimal control theory. It should be fairly obvious to the reader why the term bang-bang control is adequate.

optimal extraction rule can be written

$$h(t) = \begin{cases} 0 & \text{when } p(t)\,\mathrm{e}^{-rt} < \lambda \\ h_{\max} & \text{when } p(t)\,\mathrm{e}^{-rt} \geqslant \lambda \end{cases}$$

where $\lambda = p(T)\,\mathrm{e}^{-rT}$ is the optimal present value of a unit of the resource at the endogenously determined planning horizon T.

If the resource is extracted during a single interval $[T_0, T]$, then this interval must fulfil the condition

$$\int_{T_0}^{T} h_{\max}\,\mathrm{d}t = h_{\max}(T - T_0) = \bar{x} \tag{2.18}$$

Equation (2.18) says that the sum of the extracted resource equals the initial stock. If $h_{\max}$ approaches infinity then the interval $[T_0, T]$ shrinks to a single point (T^*), which was the optimal inventory policy summed up in proposition 2.1.

We can illustrate the extraction rule in proposition 2.3 by a diagrammatic exposition (figure 2.3), using the fact that it takes at

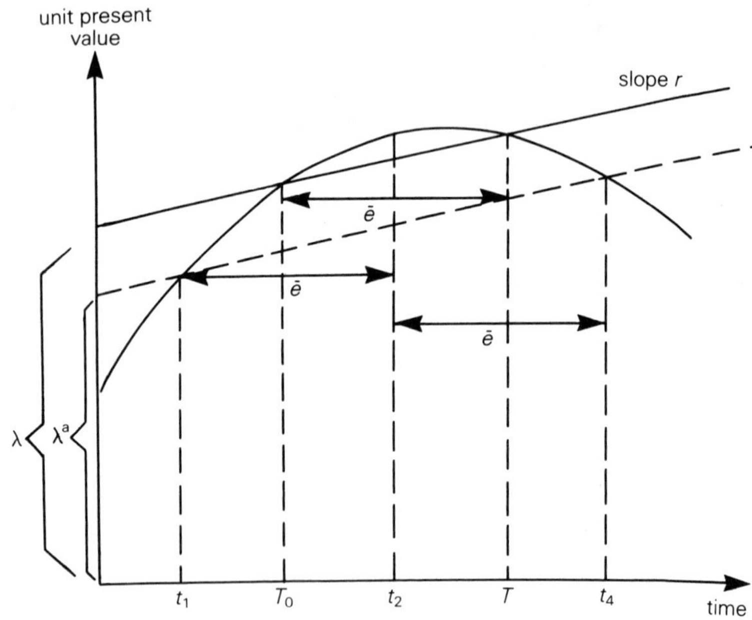

FIGURE 2.3 The determination of the optimal extraction policy when extraction capacity is limited

least time $T - T_0 = \bar{e} = \bar{x}/h_{max}$ to extract the whole stock. Since the stock is homogeneous, we can let $\bar{x} = 1$ ($\log \bar{x} = 0$). The distance between T_0 and T equals $\bar{e}$. Since the objective is to obtain the maximum present value of the resource, it is not optimal to start exploiting at $t = t_1$ and end at $t = t_2$. Nor is it profitable to delay extraction until t_2 and stop beyond T at t_4. Both alternatives imply a present value, λ^a, below λ.

2.3.3 *The Demand Function*

Let us now return to the case when extraction capacity is unlimited, i.e. $0 \leqslant h(t) \leqslant \infty$, but assume that the net price depends on the continuously differentiable demand function

$$p(t) = f[h(t)] \qquad f'[h(t)] = \frac{dp}{dh} \leqslant 0 \tag{2.19}$$

Thus function says that the net price at any t depends on the amount supplied. The greater the supply, the lower the market clearing price $[f'(h) < 0]$. Assume that the firm has perfect knowledge of its demand function and, as a result, adjusts its extraction policy to clear the market at each time.

It was shown earlier that equalizing the present value of a unit of the resource, across all extraction periods, constitutes an optimal policy under unlimited extraction capacity. Substituting the market clearing condition $p(t) = f[h(t)]$ into (2.16) yields

$$f[h(t)] = k\, e^{rt} \tag{2.16a}$$

and solving for $h(t)$,

$$h(t) = f^{-1}(k\, e^{rt}) = f^{-1}[p(t)] \tag{2.20}$$

As the net price grows at the rate $\dot{p}(t)/p(t) = r$, the supply will gradually decrease towards zero. If the net price function cuts the ordinate at a finite price, extraction declines to zero in finite time. This also means that the stock is exhausted in finite time, since it cannot be optimal to save positive amounts of the resource for which you can obtain a positive net price. Hence, another of Hotelling's original results have been 'derived' intuitively:

Proposition 2.4 (Hotelling's exhaustion rule)
If a resource stock $\bar{x}$ could be exhausted by the competitive firm subject to (2.1)–(2.5) at a rate $0 \leqslant h(t) < \infty$, and extraction takes place during an interval (T_0, T) at market clearing

prices $p(t) = f[h(t)]$, then the resource will be exhausted in finite time provided that $p = f(0)$ is finite.

To be more explicit, assume, like Hotelling, that the demand function is linear:

$$p_t = a - bh_t \qquad a, b > 0 \tag{2.21}$$

According to proposition 2.2, along the extraction path

$$\dot{p} = rp(t)$$

or

$$p(t) = p_0 \, e^{rt} \tag{2.22}$$

where $p_0 = k$ is the price level at the initial date. By solving (2.21) for h_t and substituting for $p(t)$ from (2.22), one obtains

$$h(t) = \frac{a}{b} - \frac{p_0}{b} e^{rt} \quad \text{and} \quad h_t = 0 \text{ for } p_t \geqslant a \tag{2.20a}$$

Hence, the exhaustion date T and the initial price p_0 are determined by the two equations

$$\begin{aligned} &p_0 \, e^{rT} = a \\ &\int_0^T \left(\frac{a}{b} - \frac{p_0}{b} e^{rt} \right) dt = \frac{aT}{b} + \frac{p_0}{br} (1 - e^{rT}) = \bar{x} \end{aligned} \tag{2.23}$$

2.3.4 *The Socially Optimal Extraction Path*

Is the extraction path specified in propositions 2.2 and 2.4 socially optimal? To answer this, consider a one household economy[11] with a utility function for the non-renewable resource given by

$$\begin{aligned} &U = U[h(t)] \\ &\frac{dU}{dh} = U_h [h(t)] > 0 \end{aligned} \tag{2.24}$$

Equation (2.24) means that utility is an increasing function of the consumption of the non-renewable resource. It also means that utility at a particular time is unaffected by extraction at any other time. Thus intertemporal utilities may be summed to obtain total

[11] This assumption means, among other things, that we disregard distributional consequences.

utility. A continuous version of the household's maximization problem can therefore be written

$$\max_{h(t)} \int_0^\infty U[h(t)]\,\mathrm{e}^{-rt}\,\mathrm{d}t$$

subject to

$$\dot{x}(t) = -h(t) \qquad \bar{x} = x(0) \tag{2.25}$$

and

$$\int_0^\infty h(t)\,\mathrm{d}t = \bar{x}$$

For the present value of utility to be maximized over an interval (T_0, T) during which extraction takes place, the present value of the last unit consumed (the marginal utility U_h) must be constant and equal over all periods. Otherwise it would be profitable to increase consumption in periods when the present value of the last unit consumed is high, and decrease consumption during periods when the present value of the last unit consumed in low. Formally, this rule can be written

$$U_h[h(t)]\,\mathrm{e}^{-rt} = k'$$

or (2.26)

$$U_h[h(t)] = k'\,\mathrm{e}^{rt}$$

One interpretation of (2.26) is that the marginal utility of the consumed resource increases over time at the same rate as the interest rate. Solving (2.26) for $h(t)$ yields consumer demand along thc optimal path

$$h(t) = U_h^{-1}(k'\,\mathrm{e}^{rt}) \tag{2.27}$$

where $k'\,\mathrm{e}^{rt}$ is the marginal valuation of the resource along the optimal path, i.e. the (bidding) price of the resource. For supply to equal demand along the present value maximizing path, it must be the case that $k = k' = p_0$. This means that the extraction path defined by (2.20) is socially optimal.

Proposition 2.5
The market clearing extraction rule (2.20) is socially optimal, i.e. it maximizes the present value of the sum of utilities over the planning horizon in the problem specified by (2.25).

Proposition 2.5 is not valid if non-renewable resource extraction affects the production of other commodities. In the simple model economy outlined so far, there are no extraneous commodities, firms or consumers. The presence of outside consumers who do not consume the particular non-renewable resource, but do attach a positive value to its stock, also invalidates proposition 2.5. Then the marginal valuation of the resource consumption, as expressed by the demand function, exaggerates society's true value of the units consumed. Consumers who attach positive value to the remaining stock would be willing to pay the extracting firm an amount no greater than the value of the marginal utility lost when the last unit is extracted, if the firm refrained from the extraction of this unit. Similar reasoning can be applied to the case of a firm whose production would be constrained by the extraction of the last unit of the resource. These situations are called external economies, and it is a well-known fact that external economies may distort the market economy's efficiency and optimality. This distortion also holds when external economies are present in the extraction of non-renewable resources.

The conditions for an efficient intertemporal competitive equilibrium are very restrictive. In order to demonstrate this, assume that a momentary equilibrium is characterized by (a) the clearance of the resource market at each time, and (b) a price path which follows the exponential path implied in proposition 2.2. Thus, in a momentary equilibrium,

$$\begin{aligned} h(t) &= f^{-1}[p(t)] \\ p(t) &= k\,\mathrm{e}^{rt} \end{aligned} \tag{2.28}$$

at each time. There is, however, no guarantee that this path of momentary equilibrium is intertemporally efficient. To see this, note that the only way to ensure that the resource is used efficiently is to determine k so that the resource stock is entirely exhausted at T. That is, solve

$$\int_0^T f^{-1}(k\,\mathrm{e}^{rt})\,\mathrm{d}t = \bar{x}$$

for $k = p_0$. Equations (2.28) are necessary but not sufficient for efficiency. The reason is that there is nothing in the conditions of a momentary equilibrium to ensure that the initial price is set at the correct level p_0. If $p_0' > p_0$ extraction will be too conservative, which means that consumption may be increased at every point in

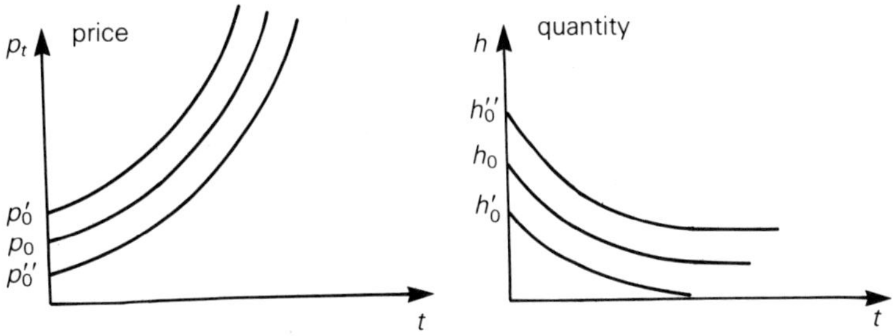

FIGURE 2.4 Efficient and inefficient momentary equilibria

time. If $p_0'' < p_0$, the resource will be extracted at too fast a rate (see figure 2.4).

The only way to guarantee the efficiency of the market solution is to provide a complete set of forward markets, which are cleared at $t = 0$ for the entire future.[12]

2.4 A More Formal Analysis

In this section we use a more formal approach to the competitive exploitation problem. The mathematical method used is called optimal control theory, and, roughly speaking, is a generalization of the classical calculus of variation.[13]

The model used in the following differs in only one respect from the model used in sections 2.2 and 2.3. We now introduce the cost function explicitly:

$$\begin{aligned} C &= C(h) \\ C_h(h) &= \frac{dC}{dh} > 0 \qquad (2.29) \\ C_{hh}(h) &= \frac{d^2C}{dh^2} > 0 \end{aligned}$$

[12] Malinvaud (1953) was the first to recognize and analyse the possible inefficiency of an intertemporal competitive economy with no terminal date. See also Dasgupta and Heal (1979).

[13] A good introduction is provided by, e.g. Intriligator (1971), chapter 12. See also Sydsaeter (1981).

Equation (2.29) means that exploitation costs depend on the rate of exploitation h, and that marginal exploitation costs are positive ($C_h > 0$) and increase with the rate of exploitation ($C_{hh} > 0$). In mathematical terminology, assumptions (2.29) mean that the cost function is strictly convex, i.e. for any two points h', h'', and $0 < \lambda < 1$, it holds that[14]

$$\lambda C(h') + (1-\lambda)\, C(h'') > C[\lambda h' + (1-\lambda)\, h''] \tag{2.30}$$

We also assume that there are N firms in the extraction industry. Among other things, this means that the production of each firm must be expressed by index i. The management problem of each firm may be formulated as

$$\max_{h_i,\, T_i} PV_i = \int_0^{T_i} [ph_i - C(h_i)]\, \mathrm{e}^{-rt}\, \mathrm{d}t \tag{2.31}$$

subject to

$$\dot{x}_i = -h_i$$

$$x_i(0) = \bar{x}_i$$

$$\int_0^{T_i} h_i(t)\, \mathrm{d}t = \bar{x}_i$$

$$0 \leqslant h_i \leqslant \infty$$

The variables h_i and T_i are called control variables, while x_i is a state variable. To solve this problem by control theory, formulate the Hamiltonian

$$H = [ph_i - C(h_i)]\, \mathrm{e}^{-rt} - \lambda_i h_i \tag{2.32}$$

According to the maximum principle, the optimal rate of extraction must maximize H over the control range $0 \leqslant h_i \leqslant \infty$ for each $t \geqslant 0$. An 'interior' solution to this maximization problem must fulfil

$$\frac{\partial H}{\partial h_i} = (p - C_{h_i})\, \mathrm{e}^{-rt} - \lambda_i = 0 \tag{2.33a}$$

and

$$\dot{\lambda}_i = -\frac{\partial H}{\partial x_i} \tag{2.33b}$$

[14] If the inequality is reversed the function is strictly concave. The analysis in this section has partly been borrowed from Clark (1976).

Since $\partial H/\partial x_i = 0$, one immediately obtains

$$\lambda_i = \text{constant} \tag{2.34}$$

The reader may also have noted that we have formulated the maximization problem in such a manner that the time horizon and the rate of extraction are control variables. This assumption imposes the following condition on the optimal path:

$$H[x_i(T_i), h_i(T_i), \lambda_i(T_i), T_i] = 0$$

or, more explicitly,

$$\{p_i(T_i)\, h_i(T_i) - C_i[h_i(T_i)]\}\ \mathrm{e}^{-rT_i} - \lambda_i(T_i)\, h_i(T_i) = 0 \tag{2.35}$$

Finally, it must hold that

$$\lambda_i(t) \geqslant 0$$

and

$$\int_0^{T_i} h_i(t)\, \mathrm{d}t = \bar{x}_i \tag{2.36}$$

One way to solve this problem is to solve (2.33) with respect to $h_i(t)$, treating λ_i as a parameter. This solution is substituted into (2.35) and (2.36) in order to determine λ_i and T_i.[15]

The interpretation of λ_i compares with the interpretations from the more intuitive reasoning developed in sections 2.2 and 2.3. From (2.33) we see that λ_i is the present value of the firm's marginal profitability at time t. From (2.34) it follows that this amount must remain constant over the extraction period, which shows that the intuitive reasoning is correct. The constancy of marginal profitability means that firms perceive that profit cannot be increased by extracting today rather than decreasing extraction tomorrow or vice versa.

Hotelling's rule of extraction (proposition 2.2) will also hold within the more general model with explicit extraction costs, provided that $\dot{p}/p$ is interpreted as the rate of change of marginal profit.

[15] There are two equations in the variables λ_i and T_i:

$$\{p_i h_i(T_i; \lambda_i) - C_i[h_i(T_i; \lambda_i)]\}\ \mathrm{e}^{-rT_i} - \lambda_i h_i(T_i; \lambda_i) = 0$$

$$\int_0^{T_i} h_i(t; \lambda_i)\, \mathrm{d}t = \bar{x}_i$$

A market equilibrium in the N identical firm competitive market model for the non-renewable resource fulfils

$$
\begin{aligned}
&\text{(a)}\ (p - C_{h_i})\,\mathrm{e}^{-rt} = \lambda_i \qquad \lambda_i = \text{constant for all } i\\
&\text{(b)}\ \{p(T_i)\,h_i(T_i) - C_i[h_i(T_i)]\}\,\mathrm{e}^{-rT_i} = \lambda_i h_i(T_i) \quad \text{all } i\\
&\text{(c)}\ \int_0^{T_i} h_i(t)\,\mathrm{d}t = \bar{x}_i \qquad \text{all } i\\
&\text{(d)}\ p = f\left[\sum_{i=1}^{N} h_i(t)\right]
\end{aligned}
\tag{2.37}
$$

where (2.37d) is the market clearing condition. This condition shows that production in a single firm affects the market price, although we here assume that it acts as if it did not.

2.5 Monopoly

A competitive firm is a firm that acts as if the market price is independent of its actions. This behaviour is reasonable if the firm is small in comparison with the market,[16] i.e.

$$h_i \ll \sum_{i=1}^{N} h_i$$

A monopoly, on the other hand, recognizes that it can affect the market price by its output decisions. An interesting question is whether a monopolist will exhaust the resource at a faster or a slower rate than a competitive industry. From the static theory of monopoly and perfect competition, we know that monopoly output is lower than competitive output. It is therefore a reasonable conjecture that, compared with a competitive industry, a monopolist will initially restrict output and raise price. Price and output paths over time would then resemble those in figure 2.5, which have been borrowed from Devarajan and Fisher (1981).

A rigorous analysis shows that these intuitive conclusions are likely outcomes, but not the only outcomes. In particular it has been shown by, among others, Lewis (1976) and Dasgupta and Heal (1979) that if the price elasticity[17] of the demand curve decreases

[16] Note that a firm might *behave* as a competitive firm, even if it holds a considerable part of the market. This possibility was used in the inuitive reasoning presented in section 2.3.
[17] The price elasticity is defined as $-(\mathrm{d}\log h)/(\mathrm{d}\log p) = (\mathrm{d}h/\mathrm{d}p)(p/h)$.

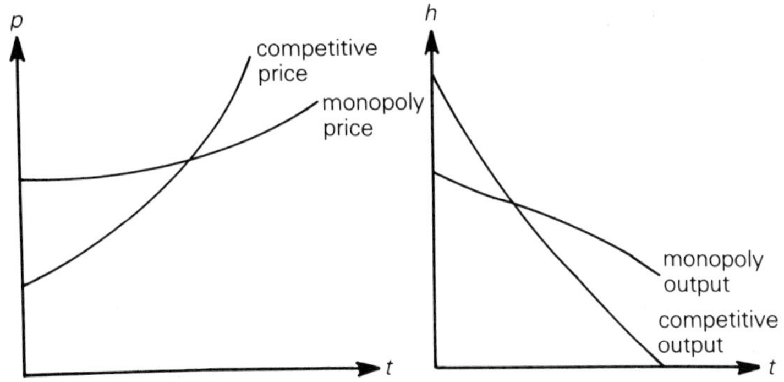

FIGURE 2.5 Competitive and monopoly price and output paths

with output, then the perfectly competitive extraction paths will resemble those in figure 2.5. We refrain from proving this claim, but note that Hotelling used a linear demand curve in his numerical example, which clearly fulfils the condition of decreasing price elasticity: the elasticity of a linear curve runs from $+\infty (h = 0)$ to zero $(p = 0)$. However, we will briefly discuss the special case of a constant demand elasticity (an iso-elastic demand curve) in order to show that monopoly under dynamic conditions may, at times, generate competitive solutions.

We begin this discussion by illustrating a few important points derived from the formal analysis of the optimal extraction policy for a monopoly firm facing a linear demand curve

$$p = a - bh \qquad a > 0,\ b > 0 \qquad [p = f(h)] \tag{2.38}$$

The reader is invited to carry out the analogous analysis for a competitive firm by using (2.33)–(2.36). Disregarding the cost function, the monopolist's maximization problem can be written

$$\max_{h,T} \int_0^T (ah - bh^2)\, e^{-rt}\, dt \qquad \left[\int_0^T f(h)\, h\, e^{-rt}\, dt\right] \tag{2.39}$$

subject to

$$h(t) \geqslant 0$$

$$\dot{x} = -h(t)$$

$$\int_0^T h(t)\, dt = \bar{x}$$

In terms of the Hamiltonian,

$$H = (ah - bh^2)\,\mathrm{e}^{-rt} - \lambda h \qquad \{H = f(h)\,h\,\mathrm{e}^{-rt} - \lambda h\} \tag{2.40}$$

Since $-\partial H/\partial x = 0$,

$$\dot{\lambda} = -\frac{\partial H}{\partial x} = 0 \tag{2.41}$$

which implies that $\lambda =$ constant. To maximize the Hamiltonian with respect to h, differentiate with respect to h and obtain the necessary condition for a maximum

$$\frac{\partial H}{\partial h} = (a - 2bh)\,\mathrm{e}^{-rt} - \lambda = 0 \quad \{[f'(h)\,h + f(h)]\,\mathrm{e}^{-rt} - \lambda = 0\} \tag{2.42}$$

where $(a - 2bh)\,\mathrm{e}^{-rt}$ is interpreted as the present value of the marginal revenue. (In the competitive case, marginal revenue is identical to price.) Equation (2.42) means that marginal revenue along the optimal path must increase at a rate equal to the interest rate. In other words, Hotelling's rule stated in proposition 2.2 holds under monopoly, when marginal revenue is substituted for (net) price.

If the resource is exhausted at some future date T, then

$$\lambda = [a - 2bh(T)]\,\mathrm{e}^{-rT} \tag{2.43}$$

or

$$\lambda = MR[h(t)]\,\mathrm{e}^{-rT}$$

where

$$MR(h) = f'(h)\,h + f(h)$$

From the maximum principle, it also follows that

$$\begin{aligned} &\mathrm{e}^{-rT}[ah(T) - bh(T)^2] - \lambda h(T) = 0 \\ &\{H[h(T), x(T), \lambda(T), T] = 0\} \end{aligned} \tag{2.44}$$

This condition is fulfilled when[18]

$$\lambda = (a - bh)\,\mathrm{e}^{-rT} \tag{2.45}$$

Combining (2.45) and (2.43), we obtain

$$\lambda = [a - bh(T)]\,\mathrm{e}^{-rT} = [a - 2bh(T)]\,\mathrm{e}^{-rT} = \lambda \tag{2.46}$$

[18] The condition can also be fulfilled if $h(T) = 0$, but in that case one obviously gets the same answer.

which implies $h(T) = 0$ and $\lambda = a\,\mathrm{e}^{-rT}$, where a is the ordinate at the origin. Substitution of (2.45) into (2.42) gives

$$h(t) = \frac{a}{2b}\,[1 - \mathrm{e}^{r(t-T)}] \tag{2.47}$$

and the time horizon is determined from the equation

$$\int_0^T \frac{a}{2b}\,[1 - \mathrm{e}^{r(t-T)}]\,\mathrm{d}t = \bar{x} \quad \left[\int_0^T h(t)\,\mathrm{d}t = \bar{x}\right] \tag{2.48}$$

which can be shown to have a unique solution. Finally, the market price of the resource is obtained as

$$p(t) = \frac{a}{2}\,[1 + \mathrm{e}^{r(t-T)}] \qquad \{p(t) = f[h(t)]\} \tag{2.49}$$

Consider now another special case, the iso-elastic demand curve $p = ch^{-\alpha}f(t)$, with $0 < \alpha < 1$. The price elasticity is $\alpha^{-1} > 1$, and the counterpart of (2.43) is

$$ch^{-\alpha}f(t)(1-\alpha)\,\mathrm{e}^{-rt} = \lambda$$

or $\hfill$ (2.43a)

$$MR_t = p(t)(1-\alpha) = \lambda\,\mathrm{e}^{rt}$$

Hence, along the optimal path, it must hold that[19]

$$\frac{\dot{p}}{p} = r$$

and $\hfill$ (2.50)

$$\int_0^\infty f^{-1}[p(t)]\,\mathrm{d}t = \bar{x}$$

and the solution is identical with the competitive solution. Readers familiar with the static theory of monopoly may find this result a bit surprising. It should, however, be remembered that we are considering monopoly power over a stock and a flow, not only a flow as in the traditional theory of monopoly. Keeping this observation in mind, it should be a bit easier to grasp the fact that a monopoly may extract the resource at a faster rate than under competitive conditions. This result holds when there are no extraction costs,

[19] Note that $0 < \alpha < 1$ is necessary as otherwise $MR_t < 0$ for all $p > 0$.

provided that the price elasticity of demand decreases with the amount.

2.6 Summary

A common feature of the extraction problems which have been analysed in this chapter is that we can restate proposition 2.2 in a way that makes it valid for all optimal solution paths. The underlying principles that make this possible were given in connection with the intuitive reasoning in section 2.3. The *shadow price* λ takes on different values in the different problems that have been considered. However, this shadow price (constant in all the solutions considered) can be interpreted as the maximal amount the resource owner would be willing to pay for the option to extract one unit of the resource at some future date. This means that λ is the present value of the profit per unit of output that can be earned along the optimal path. Since there are $\bar{x}$ units in total, the present value of the profits from extraction is $\lambda\bar{x} = K$, which is the value of the capital owned by the extracting firm. Every time a unit of the resource is extracted, the present value of the capital is reduced by a constant amount λ. This amount is constant over time because the amount left, $x(t)$, does not affect revenues or costs. Clearly, if extraction costs were a decreasing function of the stock,[20] a unit of the resource at time zero would be expected to have a higher present value along the optimal path than a unit of the resource at some future time.

Since the present value of a unit of the resource along the optimal path is constant over time, the current value of a unit of the resource must increase at the same rate as the interest rate, i.e. $\dot{\lambda}'/\lambda' = r$ where $\lambda' = \lambda\, e^{rt}$. This simple relationship does not hold if extraction costs depend on the amount extracted, and might be one reason why the so-called Hotelling's rule is not easy to discover in the empirical evolution of profits in mining, for example.

Finally the reader should note an important deviation between the static theory of perfect competition[21] and the theory of non-renewable and renewable resources under competitive conditions. In static price theory, supply is directly related to marginal cost, i.e. the cost of the last unit produced, while no simple relation between

[20] This condition is very likely to be fulfilled in mining, for example, as you have to dig deeper and deeper to extract additional units of the ore.

[21] For the static theory of perfect competition consult any elementary textbook in price theory.

marginal production costs and supply exists in the dynamic theory of resource management. The confused reader is asked to return to Hotelling's rule for an illustration of this proposition. This phenomenon is not particular to non-renewable resources, but will hold for most of the results on supply decisions that will be derived in the chapters to follow. The curious reader is referred to chapter 11 on the rules of costs and benefits.

3

The Exploitation of Renewable Resources

3.1 Introduction

The purpose of this chapter is to review the standard theory of renewable resources. A widely held view is that this theory has a very broad spectrum of applications. Some of these are the cutting of trees, the management of fisheries and the killing of game animals.[1] We feel that many such applications of the theory are often exaggerated. The biological properties of the models presented in this chapter are perhaps best suited to analyse the 'harvesting of bacteria',[2] but will also give a good description of the problems involved in the catching of fish, and will provide some insights into the management problem involved in biomass production. As we show later, the model is not well suited to analyse all of the economic aspects of forestry.

The basic framework of renewable resource models is quite simple. A (fish) population, or biomass, is viewed as a capital stock, in that it is capable of yielding a sustainable consumption flow over time. The economic problem consists of choosing an optimal consumption flow through time. This choice yields the optimal stock level at every time.

The pioneering work in this field is a paper by Scott (1955), which casts the fishery problem in capital theoretic terms. This paper was followed by Crutchfield and Zellner (1962), where the fishery problem is formulated in terms of a dynamic model. However, as Clark and Munro (1975) point out: 'In spite of these works, however, the received theory of fisheries economics founded by

[1] For 'optimistic' views like this, see Beddington, Watts and Wright (1975).
[2] This expression is taken from Borck (1976)

Gordon (1954) continued to be formulated in static terms.' The main reason was probably the lack of adequate mathematical instruments within capital theory. The dynamic optimization method prior to Pontryagin et al. (1962) was the calculus of variations. This method, although applicable in principle, proved to be inadequate. Dorfman (1969) argues that modern capital theory was born with optimal control theory, and it is clear that this new optimization technique has had a great impact on the development of resource economics as well.

In section 3.2 different variations of the modern theory of renewable resources are reviewed.[3] In section 3.3, we discuss the so-called problem of the commons. We see that a free access fishery is likely to generate inefficiencies since the resources devoted to fishing are too large. We end the chapter by introducing a land use conflict problem where forestry is involved (forestry versus reindeer husbandry). This problem is used to illustrate what is referred to as Coase's theorem.

As in the previous chapter, some parts of the presentation will be fairly technical and demand some knowledge of control theory. However, we have tried to phrase the interpretation of the results in a common sense language which should be grasped by all readers.

3.2 The Bang-bang Control Renewable Resource Model

The biological growth function is built on a model of population dynamics which was first presented by Schaefer (1954). Schaefer's model rests upon the so-called Pearl-Verhultst, or logistic, equation of population dynamics.[4] A general version of this explicit equation is

$$\begin{aligned} x &= x(t) = \text{the biomass stock at time } t \\ \dot{x} &= F(x) \end{aligned} \tag{3.1}$$

Equation (3.1) expresses biomass growth as a function of the biomass stock. The *growth function* $\dot{x}$ is assumed to have the following properties:

[3] There are many contributions to the modern theory of renewable resources and it is not our intention to review all of them. We would, however, like to mention the advanced textbooks by Clark (1976) and Dasgupta and Heal (1979), which have had a large impact on our presentation of the theory.

[4] The equation is $F(x) = \alpha x(1 - xK^{-1}) - \beta$, where α, β and K are positive constants. For $\beta = 0$, K can be interpreted as the carrying capacity of the environment.

$$
\begin{aligned}
&\text{(a)}\ F(x) > 0 && \text{for } x \in (x_1, x_2) \\
&\text{(b)}\ F(x_1) = F(x_2) = 0 && && (3.2) \\
&\text{(c)}\ \frac{d^2F(x)}{dx^2} = F''(x) < 0 && x \in (0, \infty)
\end{aligned}
$$

Property (a) means that the stock is growing as long as it remains within the open interval (x_1, x_2). The second property means that growth ceases at a population size of x_1 or x_2. Finally, the third property means that the marginal rate of growth is a strictly decreasing function of the population size. These assumptions generate a half circle shaped curve (see figure 3.1). The growth rate increases monotonically, $F(0) = -\beta < 0$, to some critical level of the stock $\bar{x}$, the maximum sustainable yield of the population, and then decreases monotonically. If the system is left to itself, it will end up at $x = x_2$, provided that the initial population is larger than x_1. Formally,

$$\lim_{t \to \infty} x(t) = x_2 \qquad x(0) > x_1$$

where x_2 is a stable equilibrium, which may be interpreted as the environment's carrying capacity. On the other hand, if the initial

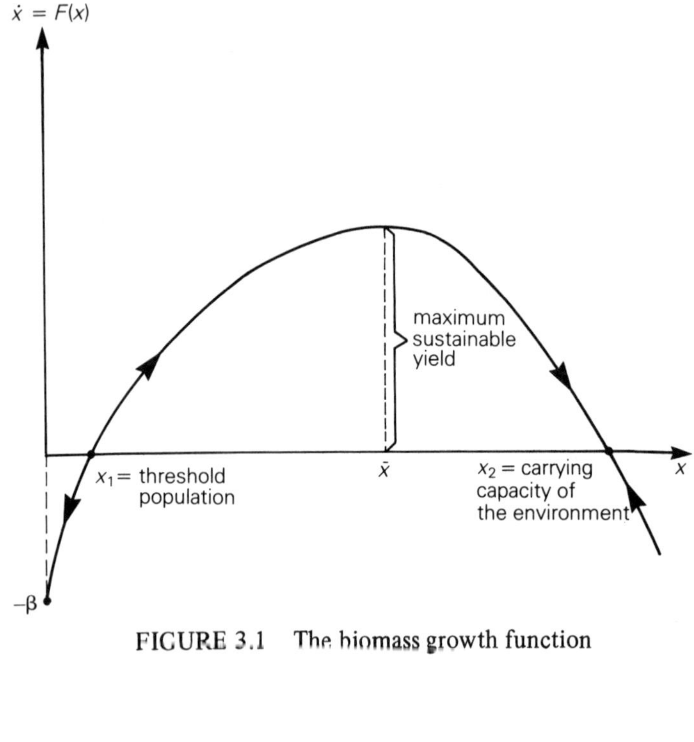

FIGURE 3.1 The biomass growth function

stock is lower than x_1 (the *threshold population*) there is no return; the species is extinguished. $x = x_1$ is an unstable equilibrium, while $x = x_2$ is a stable equilibrium.

The *production function* may be expressed more generally as

$$\dot{x} = F(x; \alpha)$$

where α is a vector of unalterable or alterable environmental characteristics. Alterable characteristics are things like fertilization, while the weather is an unalterable characteristic. By introducing *harvesting* into the model as a function of time, we can write the production function as

$$\dot{x} = F(x) - h(t) \qquad (3.3)$$

$$h(t) \geqslant 0$$

The difference between the growth rate $F(x)$ and the harvest rate $h(t)$ may be interpreted as an investment which adds to the stock of the biomass. We see that the growth function of a non-renewable resource is just a special case of (3.3), when $F(x) \equiv 0$.

We are now ready to consider the economic resource management problem in a competitive setting. Assume that the firm which controls the resource is small relative to the overall market, i.e. the firm's supply does not have any significant influence on the market price, and the current price is constant over time. Moreover, assume that the unit cost of harvesting is constant and independent of the biomass stock. Later we will introduce the more realistic assumption that the average harvesting cost is a decreasing function of the size of the biomass. It is clearly easier to catch fish when a larger stock exists, and similar conditions are probably present when the biomass consists of live trees. One also expects that the unit cost of harvesting is a function of the volume of the harvest. Hence in most instances the constant average cost assumption is probably a bit dubious.

The problem facing the profit maximizing firm is

$$\max_{h(t)} \int_0^\infty ph(t)\, \mathrm{e}^{-rt}\, \mathrm{d}t \qquad (3.4)$$

subject to

$$\dot{x} = F[x(t)] - h(t)$$

$$0 \leqslant h \leqslant h_{\max}$$

$$x(t) \geqslant 0$$

$$x(0) = x_0$$

and where p is the net profit per unit of harvest.

The restriction $0 \leqslant h \leqslant h_{max}$ is due to the firm's fixed capacity.[5] It says that one can choose the level of capacity utilization in the interval between total inactivity and full utilization.

Intuitively, according to figure 3.1, a stock equal to $\bar{x}$ generates the maximum possible sustainable harvest; given that p is independent of the stock and harvest rate, this policy yields the maximum dollar sustainable yield per unit of time. We call this catch policy a *steady state solution*. In a steady state, the solution catch (whatever it is) is equal to the natural rate of growth of the stock.

To determine the necessary conditions for steady state optimality, note that a switch from one steady state to another caused by extracting an extra unit of the resource today yields the direct net revenue $\mathrm{d}x^s$. On the other hand, this causes a loss for all future periods of the extra growth which would have been generated by this marginal unit,

$$-F'(x^s)\,\mathrm{d}x^s = -\frac{\mathrm{d}\dot{x}}{\mathrm{d}x}\,\mathrm{d}x^s$$

The profitability of the indicated policy change requires that the present value of the difference investment is non-negative, i.e.

$$\left[1 - \frac{F'(x^s)}{r}\right]\mathrm{d}x^s = \left[\frac{r - F'(x^s)}{r}\right]\mathrm{d}x^s \geqslant 0$$

where $F'(x^s)/r$ is the present value of a perpetuity yielding $F'(x^s)$. In other words, as long as $F'(x^s) > r > 0$ the stock should be increased by investing extra units in the 'fishing bank' ($\mathrm{d}x^s < 0$). If $r > F'(x^s)$ then extra money should be invested in the 'bank' at the rate of return r ($\mathrm{d}x^s > 0$). Finally, if $F'(x) = r$ the rates of returns in the fishing bank and the bank coincide and there are no obvious reasons to switch from one steady state to another. Moreover, since $F'(\bar{x}) = 0$ at the maximum sustainable yield, this stock cannot be an optimal steady state unless $r = 0$. For $r > 0$ the optimal steady state stock is smaller than the maximum sustainable yield level, since $F'(x) < 0$ for $x > \bar{x}$.

[5] To include the possibility of investment in new capacity generates a series of interesting, but difficult, analytical problems. Many of them were solved by Clark, Clark and Munro (1979).

In order to derive the solution more formally, we state the Hamiltonian of the control problem (3.4):

$$H = ph(t)\,\mathrm{e}^{-rt} + \lambda\{F[x(t)] - h(t)\} = S(t)\,h(t) + \lambda F(x) \quad (3.5)$$

where

$$S(t) = p\,\mathrm{e}^{-rt} - \lambda$$

is the *switching function*.

The Hamiltonian is linear in the control variable $h(t)$. Pontryagin's optimal control theory demands that $h(t)$ be chosen to maximize H. Thus it is optimal to choose $h(t) = h_{\max}$ when $S(t) > 0$, and $h(t) = 0$ when $S(t) < 0$. In addition, an optimal solution must obey the 'law'

$$\dot{x} = F(x) - h$$

and it must also hold that

$$\text{(a)}\ \dot{\lambda} = \frac{-\partial H}{\partial x} = -\lambda F'(x)$$

$$\text{(b)}\ \lim_{t \to \infty} \lambda x = 0 \qquad \text{(transversality condition)} \quad (3.6)$$

To preserve $S(t) = 0$, we must choose h in a certain manner. For $S(t) = 0$ on some interval (t, ∞) it must hold that

$$\frac{\mathrm{d}S}{\mathrm{d}t} = -(rp\,\mathrm{e}^{-rt} + \dot{\lambda}) = 0 \quad (3.7)$$

Using (3.6) and $S(t) = 0$, one obtains

$$r\lambda = \lambda F'(x') \quad (3.8)$$

or

$$r = F'(x')$$

Hence, the steady state solution is given by the equality between the interest rate and the yield change that results from increasing the stock by one unit (the increased consumption possibilities which are created from an increase in the population of one unit). As seen in figure 3.2, this solution results in biological 'overfishing' since the steady state stock x' is lower than the maximum sustainable yield stock $\bar{x}$.

We now summarize the management rule for the problem specified by (3.4):

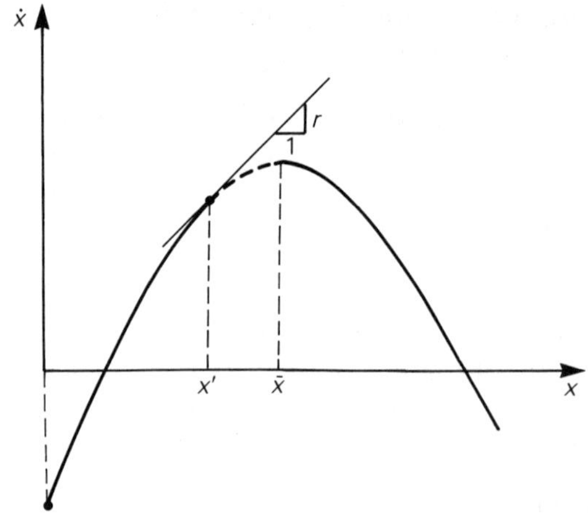

FIGURE 3.2 The steady state for model (3.4)

Proposition 3.1
The optimal policy for the model specified in (3.4) is

(a) If $x(0) > x'$, set $h = h_{max}$ and deplete the stock until $x = x'$ for some t''. Then set $h = F(x')$ for $t \in (t'', \infty)$.
(b) If $x(0) < x'$, set $h = 0$ and let the stock grow until $x = x'$ for some t'. Then set $h = F(x')$ for $t \in (t', \infty)$.

The optimal solution is to deplete (augment) the resource, using bang-bang control, to its optimal steady state level and then preserve it at this level by harvesting the steady state growth. In other words, 'invest' or 'consume' until the stock reaches an optimal level and afterwards follow a non-investment path which keeps the optimal capital stock intact[6] (see figure 3.3).

The reason why bang-bang control is optimal is that the harvest level does not affect the unit cost of harvesting directly; there is only a change in the opportunity cost of not harvesting through the change in the stock. This suggests that the optimal rule would be changed if we assume that unit cost is an increasing function of harvest quantity. In this case the resource should be depleted towards its optimal steady state level by means of an adjustment process smoother than bang-bang control.

[6] This solution fulfils the transversality condition $\lim_{t\to\infty} \lambda x = 0$, since $\lambda = p\,e^{-rt}$ and $x = x' =$ constant for some $t > 0$.

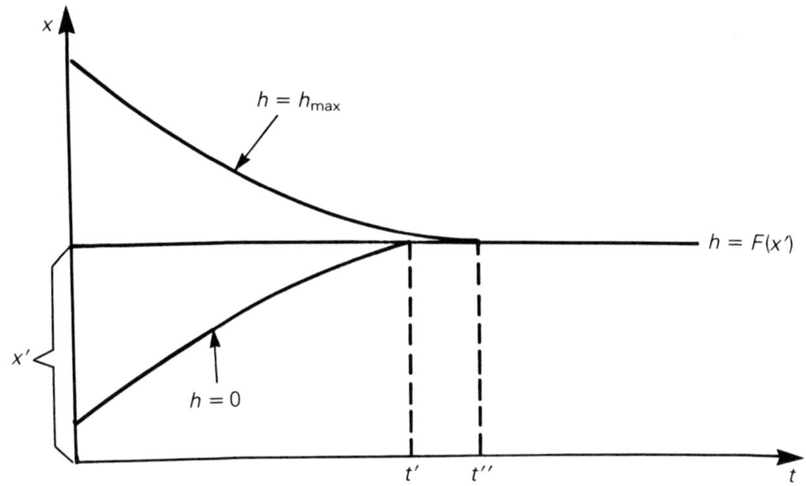

FIGURE 3.3 The optimal harvest policy when the Hamiltonian is linear in the control variable h, the rate of harvest

On the other hand, if unit harvesting costs are a function of the stock, but not a function of the rate of harvest, then the optimal harvesting rule can be described in a manner analogous to the rule specified by proposition 3.1. This holds because the Hamiltonian remains linear in the control variable h.[7] However, the resulting steady state does not necessarily result in overfishing, in the sense that the steady state stock is smaller than the maximum sustainable yield stock. The resulting steady state stock may well be larger than the maximum sustainable yield stock. If we assume that unit harvesting costs decrease with the size of the stock $[\mathrm{d}c(x)/\mathrm{d}x = c'(x) < 0]$, a necessary condition for overfishing is

$$p - c(\bar{x}) > 0 \qquad \text{(unit profit)}$$

where $\bar{x}$ is the population yielding maximum sustainable yield. This is seen by assuming that unit profits are negative at $\bar{x}$. Since $c'(\bar{x}) < 0$, unit profit remains negative for every additional decrease of the stock which proves the above observation; recall that p is constant and $c(x)$ increases. The reader is referred to Clark and Munro (1975)

[7] This is easily seen by rewriting the Hamiltonian as

$$H = [ph(t) - c(x)\,h(t)]\,\mathrm{e}^{-rt} + \lambda[F(x) - h(t)] = S(t)\,h(t) + \lambda F(x)$$

where $S(t) = (p - c)\,\mathrm{e}^{-rt} - \lambda$ is the switching function.

for a comprehensive analysis of the conditions under which overfishing or underfishing occurs.

We now return to the simple model in (3.4) for some comparative statics. From the steady state condition $[r = F'(x')]$ it is evident that the steady state stock x' coincides with the maximum sustainable yield stock, provided that $r = 0$. In fact, given the shape of the growth function, $r = 0$ is the only instance when $x' = \bar{x}$ in steady state. Moreover, if r increases, the steady state population is a decreasing function of the interest rate. If the interest rate rises sufficiently, we will eventually reach a steady state population which is smaller than the threshold population. This amounts to saying that a private enterprise finds it profitable to extinguish the species, as seen in figure 3.4 (see also Hoel, 1978).

The phenomenon just described is one explanation as to why the blue whale is close to extinction. A second reason, and a more clear-cut external diseconomy, will be discussed in connection with the so-called problem of the commons (see section 3.3). See also the appendix to this chapter, where the model is extended to include the case when the resource stock as such has a value to society.

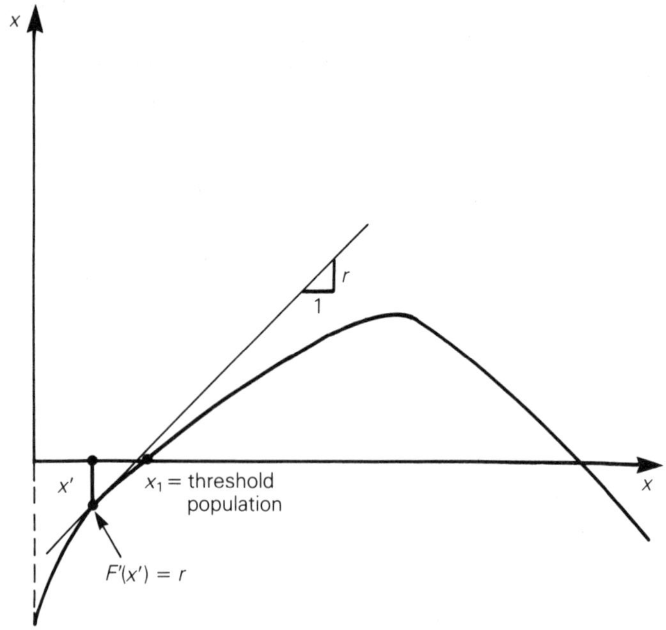

FIGURE 3.4 Extinction when steady state population is smaller than the threshold population

Extinction due to overexploitation is the conventional view of the threat to endangered species of animals and plants. A more severe threat is probably habitat modification, which can take many forms: direct conversion, as in drainage of wetlands or in harvest of tropical forests; chemical pollution, as from acid rain; and biological pollution, the introduction of exotic species. It is difficult to predict what values – a cure for cancer, a perennial corn or a fast growing pine – will ultimately emerge if the species that may produce them are not lost. Irreversibility in combination with uncertainty introduce a new element into cost–benefit analysis. There is a gain from being able to learn about future benefits that would be precluded by development if one abstains from developing initially – an option value. For a recent informative theoretical analysis of endangered species and the economics of irreversible damage, the reader is referred to Fisher and Hanemann (1983).

One interpretation of the model in (3.4) in terms of forestry management is as follows. If the biomass of growing trees is larger than the optimal stock (i.e. there are trees older than the optimal rotation age) the biomass should be cut at the maximum speed,[8] while if the biomass is too small one should wait until it has reached the optimal level. When this optimal level is reached, the yearly growth of the biomass should be harvested.

However, since there is no explicit age distribution across trees, the model does not show which trees should be cut. Thus the model cannot adequately deal with the problem which confronts forest management since it treats the biomass as homogeneous. The single biomass equation (3.1) does not distinguish between the young and old stands of equal volume, whereas the growth of the biomass in a forest is clearly a function of its age composition.

As fish are harvested, crowding occurs less and less, and under certain circumstances (population larger than the maximum sustainable yield population), the rate of growth of the remaining population may increase. This is exactly what is conveyed by the biomass equation (3.1). Trees are subject to two kinds of harvest – thinning and cutting. Thinning and other methods of stand improvement mean that only some of the trees are removed from the forest. The thinning objective is to give the remaining trees more light and hasten their growth. Clearcutting does nothing for the remaining trees since there are no remaining trees. This shows that the biomass equation (3.1) fails to provide an adequate description of growth in a forest.

[8] This management policy corresponds to the behaviour which is discussed in Streyffert (1938), pp. 153–89.

In chapters 4–6 we return to a discussion of how the growth of forest can be modelled.

The renewable resource model that we have dealt with so far obscures price dynamics. It assumes that the price level is constant over time. Readers interested in the effects of an explicit market are referred to the appendix to this chapter.

3.3 The Problem of the Commons

In the renewable resource model analysed in section 3.2, we assumed that there is a single firm which has access to the resource. This assumption is relaxed in this section. We now assume that there are *n* firms, each of which has free access to the resource.

This seemingly innocent generalization of the analysis changes the properties of the solution in a non-trivial manner. In fact, the renewable resource competitive equilibrium may well be socially inefficient, provided that the actions of a profit maximizing firm affect the output (harvest) of other firms. The underlying reason for this was touched upon in chapter 2 in connection with the exploitation of an oil well by two different firms. Another example of this kind of inefficiency is the so-called free access fishery model, where many fisheries have access to a common fishing ground. An analogous problem is also present when there is free access to a common hunting area, or when wood based industries have free access to a common forest.[9]

This problem is often referred to as the problem of the commons, which, in a broader perspective, is really an external effect in production. In a renewable resource context, the problem should be studied in a dynamic setting, where both the number of firms and the stock of the non-renewable resource are allowed to vary.[10] In order to highlight the underlying cause of the inefficiency, we use a mainly static analysis, where the number of firms is assumed to be constant and the size of the resource stock is only considered implicitly. This particular form of analysis is borrowed from Dasgupta and Heal (1979). The usual way to model free access fishery is to assume free entry of firms. The classical treatment of this problem is found in Gordon (1954).[11] One formal difference

[9] This was, for example, the case in Sweden during the eighteenth century. The resulting exploitation of the wood resource made the need for forest policy and/or property rights very clear. As an official state agency so adequately put it: 'The forests were used by everyone, but attended to by nobody.' See Stridsberg and Mattsson (1980).

[10] For a general equilibrium approach see Mäler (1974).

[11] For a recent textbook treatment see also Hannesson (1978).

between these two approaches is that the free entry case assumes that firms and vessels enter the industry as long as profits are positive. The number of firms in the fishing industry, however, only affects the amount of the inefficiency, not whether or not inefficiency occurs.

3.3.1 *The Free Access Fishery*

Assume a fishing ground of size F. The fishing biotechnology is described in (3.1) and figure 3.1. In the present context we will deal only with the fishing technology, treating the size of the stock as exogenous. The catching technology is given by the linear homogeneous production function[12]

$$c = f(x, F) \tag{3.9}$$

where c is the catch, F is the size of the fishing ground and x is the number of vessels.

From linear homogeneity it follows that[13]

$$Ff(x/F, 1) = f(x, F)$$

Moreover, since F is a constant, then by letting $F = 1$ we can rewrite the catching technology as

$$f(x, 1) = g(x)$$

which is assumed to have the properties

$$\begin{aligned}
&\text{(a)}\ g(0) = 0 \\
&\text{(b)}\ g_x(x) = \frac{\mathrm{d}g}{\mathrm{d}x} > 0 \\
&\text{(c)}\ g_{xx}(x) < 0 \\
&\text{(d)}\ \lim_{x \to \infty} g(x) = C
\end{aligned} \tag{3.10}$$

Assumption (3.10a) simply means that no vessels result in a zero catch; (3.10b) means that an increased number of vessels increases the catch (the marginal product of vessels is positive); and (3.10c) means that the marginal product of vessels is a decreasing function of

[12] $f(\lambda x, \lambda F) = \lambda c = \lambda f(x, F)$ for $\lambda > 0$. For a more thorough treatment of the properties of linear homogeneous production functions (constant returns to scale) see any introduction to production theory.

[13] Homogeneity implies that $\lambda f(x, F) = f(\lambda x, \lambda F)$. Now put $\lambda = 1/F$. This gives $(1/F)f(x, F) = f(x/F, 1)$ and $f(x, F) = Ff(x/F, 1)$.

the number of vessels. Finally, (3.10d) means, together with (3.10a–c), that the catch is bounded from above. The upper bound C is best thought of as the total stock of fish.

Given that the price of fish is p, and that the rental price of a vessel is r, society's management problem can be formulated as[14]

$$\max_x \pi(x; p, r) = pg(x) - rx \tag{3.11}$$

i.e. choose x – the number of vessels – to maximize industry profits. The first order condition for a maximum is

$$pg_x(x) = r \quad \left[\text{or } g_x(x) = \frac{r}{p}\right] \tag{3.12}$$

Thus, at an optimum, the marginal revenue product of vessels is equal to the rental price of vessels.[15] The solution of (3.12) is a function of the price of fish and the rental price of vessels. This can be written as

$$x^* = x^*(\underset{+}{p}, \underset{-}{r}) \tag{3.13}$$

where the signs under the expression denote the signs of the partial derivatives. These signs should be fairly obvious from figure 3.5.

Let us now consider a case where n identical firms have free access to the common fishing ground F. The catch of firm i depends on the number of vessels used by all other firms and the number of vessels used by i, i.e. the total number of vessels. This can be written

$$c_i = x_i \frac{g(x)}{x} \tag{3.14}$$

where c_i is the catch of firm i, x_i is the number of vessels used by firm i,

$$x = \sum_{i=1}^{n} x_i$$

[14] In chapters 10 and 11 it will be shown more formally that profit maximization in a competitive setting without externalities leads to a social optimum (a Pareto efficient allocation). For the moment, the suspicious reader is asked to believe our assertion that the problem in (3.11) is the 'correct' problem in the search for a social optimum.

[15] This price should be interpreted as a measure of the value of the production which is lost in the other sectors of the economy when another vessel is used for fishing purposes. The term *rental price* is used as it is implicitly assumed that vessels can be rented in a perfect market. Hence, no investments in vessels are necessary.

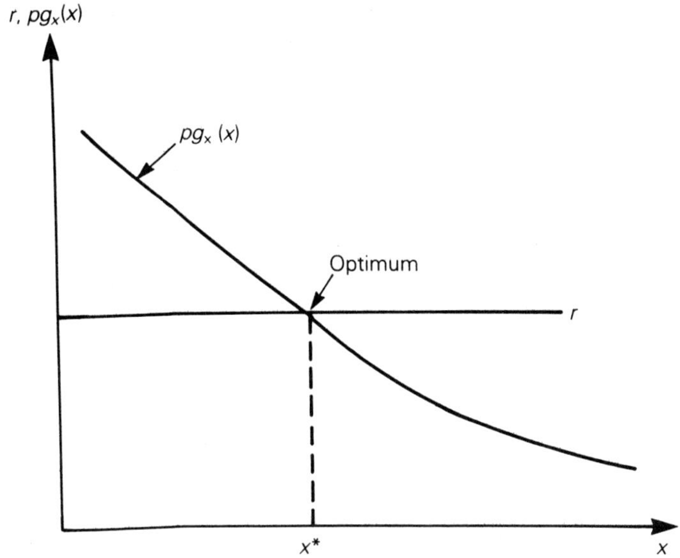

FIGURE 3.5 The socially optimal number of vessels

is the total number of vessels in the industry and $g(x)/x$ is the average catch per vessel.

Equation (3.14) states that the catch of firm i is equal to the number of vessels used by firm i times the average catch per vessel.

The profit of firm i can be written

$$\pi_i = px_i \frac{g(x)}{x} - rx_i \tag{3.15}$$

and for i to maximize profits it is necessary that

$$\frac{\partial \pi_i}{\partial x_i} = p\left[\frac{g(x)}{x} + x_i \frac{g_x(x)}{x} - x_i \frac{g(x)}{x^2}\right] - r = 0 \quad \text{all } i \tag{3.16}$$

Since all firms are assumed to be identical, $x_i = x_j = \bar{x}$ and $x = n\bar{x}$. This enables us to rewrite (3.16) as

$$(n-1)\bar{x}\frac{g(n\bar{x})}{n\bar{x}} + \bar{x}\frac{g_x(n\bar{x})}{n\bar{x}} = \frac{r}{p}$$

or (3.16a)

$$\frac{g(n\bar{x})}{n\bar{x}} - \frac{1}{n}\left[\frac{g(n\bar{x})}{n\bar{x}} - g_x(n\bar{x})\right] = \frac{r}{p}$$

The $\bar{x}$ which solves this equation is the equilibrium number of vessels per firm but, since $n\bar{x} = x$ is equilibrium,

$$\frac{g(x)}{x} - \frac{1}{n}\left[\frac{g(x)}{x} - g_x(x)\right] = \frac{r}{p} \tag{3.17}$$

Denote the solution $x^0 = n\bar{x} = x^0(p, r)$. We are interested in comparing the solution of (3.17) with the socially optimal solution obtained from (3.12), i.e. we want to compare the magnitudes of x^0 and x^*. Therefore, subtract $g_x(x^0)$ from both sides of (3.17) and rewrite the equation as

$$\frac{r}{p} - g_x(x^0) = \frac{n-1}{n}\left[\frac{g(x^0)}{x^0} - g_x(x^0)\right] \tag{3.18}$$

From assumptions (3.10) about the properties of the production functions, it follows that[16]

$$\left[\frac{g(x^0)}{x^0} - g_x(x^0)\right] > 0 \qquad \text{all } x^0$$

which means that the average catch per vessel is larger than the catch of the marginal vessel. From (3.18) it follows that the real rental price of vessels (r/p) in equilibrium is larger than the catch of the marginal vessel. In the social optimum, there is an equality between these amounts [from (3.12)]. From decreasing returns to scale $g_{xx} < 0$, it then follows that $x^0 > x^*$, or:

Proposition 3.2
Given a free access fishery with the catch technology specified by (3.10), then the total number of vessels and the total catch will be larger than the socially optimal number of vessels and the socially optimal level of the total catch.

Put differently, the profit of each firm can be increased by decreasing the total number of vessels. The rental price of the last vessel is larger than the value of the catch contributed by the last vessel [from (3.18)]. Hence, by reducing total fishing efforts, the industry's total profit can be increased, and this increase may be shared among all n firms. The 'extra' profit possibilities disappear when $r = pg_x(x)$, i.e. when (3.12) is fulfilled. Since the total catch is too large, the risk of depleting the stock below the threshold level

[16] Define $H(x) = g(x) - xg_x(x)$. Clearly $H(0) = 0$ and $H'(x) = -xg_{xx}(x) > 0$ for all $x > 0$. Hence, $H(x) > 0$ for all $x > 0$.

in figure 3.1, i.e. making the species extinct, increases under a free access fishery.

3.3.2 *Regulation of a Free Access Fishery*

The reason for market failure under a free access fishery is that a firm's introduction of an extra vessel increases its catch but, at the same time, inflicts a diseconomy on other firms by reducing their catches. An understanding of this phenomenon provides the key to solving the problem of the commons.

Assume that a single fishery is granted the property rights to the fishing ground. The firm cannot now damage other firms by its actions and this should result in a more socially efficient fishing effort. This is indeed the case, since $n = 1$ in (3.18) transforms this equation to (3.12), and $x^* = x^0$. By specifying the property rights in this manner, the externality is internalized, and the costs and revenues of the firm coincide with the costs and revenues of society as a whole. The reader should note that property rights do not necessarily result in monopoly since other firms may 'own' fishing grounds which supply the same species of fish.

A variation of this property rights method is if the fishing area is divided into n plots of equal size, and each firm is given the property rights to one plot. Provided that the property rights could be protected without cost[17] this would also yield a socially optimal catch.

These methods transform the free access problem into a management problem closely related to the management problem analysed in section 3.2. The difference is that the biotechnology and the time dimension are more explicitly considered in the previous problem.

A third solution to the problem of the commons is to use a tax scheme that forces each firm to pay a tax for each vessel it operates. Set the tax per vessel t equal to the left hand side of (3.18), evaluated at the socially optimal number of vessels x^*, i.e.

$$t = p\,\frac{n-1}{n}\left[\frac{g(x^*)}{x^*} \quad g_x(x^*)\right] \tag{3.19}$$

The first order condition for the individual firm i becomes

$$\frac{g(nx_i)}{nx_i} - \frac{1}{n}\left[\frac{g(nx_i)}{nx_i}\right] = -\,g_x(nx_i)\,\frac{r}{p} + \frac{t}{p}$$

[17] The problem is that there are incentives both to attract fish from other fishing areas and to fish in other firms' fishing areas.

or

$$\frac{n-1}{n}\left[\frac{g(nx_i)}{nx_i} - g_x(nx_i)\right] + g_x(nx_i) = \frac{r}{p} + \frac{t}{p} \tag{3.20}$$

Set $x_i = x^*/n$ for all i. Substitution of (3.19) into (3.20) then yields

$$g_x(x^*) = \frac{r}{p} \tag{3.21}$$

Thus, the tax is set equal to the difference between social and private marginal costs at the social optimum. Consider figure 3.6. MC_p is private marginal cost, which underestimates society's marginal cost MC_s. By introducing a unit tax per vessel of size t, private and social marginal costs coincide at the social optimum x^*/n, and each profit maximizing firm is induced to produce at the socially optimal rate.

A fourth and widely used method to regulate the catch of a free access fishery is to limit the total catch, or to limit the number of vessels to x^*/n, and thereby solve problems of external diseconomy. This also holds if the total catch is limited to $g(x^*)$, provided that

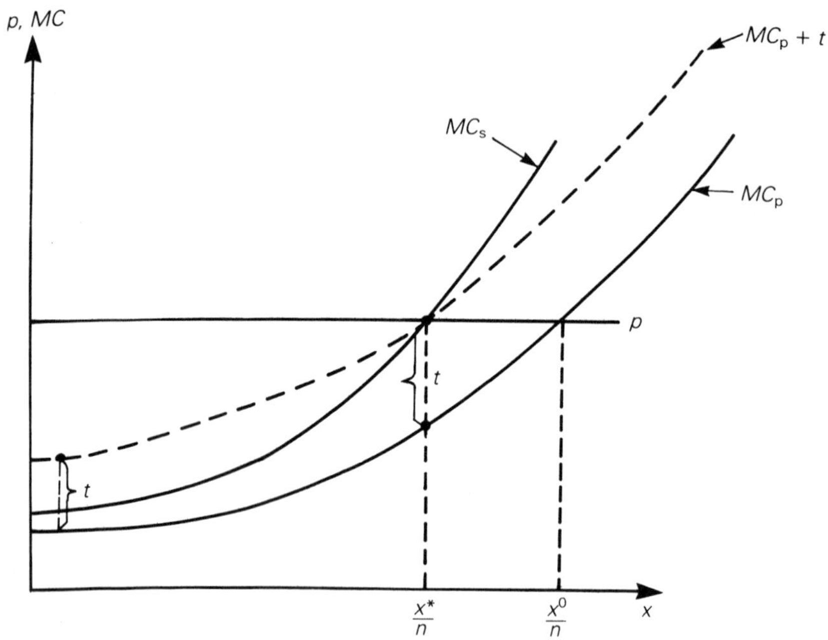

FIGURE 3.6 Tax correction of the externality

this limit can be enforced without cost, i.e. without using resources which could be used elsewhere in the economy.

Each of these methods to solve the problem of the commons has many associated practical difficulties, but we will not discuss these here. We end this chapter by introducing a land use conflict problem involving forestry which is very similar to the common property problem discussed in this section. The solution of this problem hints at a simple method which can be used to solve related problems.

3.4 Coase's Theorem: An Illustration

In the northern part of Scandinavia there is a land use conflict between reindeer husbandry and forestry. Both industries use common land. The Laplanders claim that modern forestry destroys, among other things, the pasture for the reindeer, and there are claims from forestry that the reindeer destroy the plants during the regeneration period. The land use conflict is a clear cut example of external diseconomies. The production function of forestry depends on the intensity of reindeer husbandry and the production function of reindeer husbandry depends on the intensity of forestry.

To simplify the analysis we will assume that the intensity of forestry affects reindeer husbandry negatively, while the effects of reindeer husbandry on forestry are negligible.

Let $B^{F}(x)$ be the net revenue from forestry, when forestry is conducted at intensity x. Moreover, let $B^{R}(x)$ be the net revenue that falls on reindeer husbandry when forestry is conducted at intensity x. The socially optimal scale of forestry is the solution to

$$\max_{x} S(x) = B^{F}(x) + B^{R}(x) \tag{3.22}$$

i.e. choose x such that the sum of the net revenues is maximized. If the function $S(x)$ is differentiable, we get the first order necessary condition for a socially optimal forestry scale:

$$\frac{\mathrm{d}S(x)}{\mathrm{d}x} = \frac{\mathrm{d}B^{F}(x)}{\mathrm{d}x} + \frac{\mathrm{d}B^{R}(x)}{\mathrm{d}x} = 0 \tag{3.23}$$

The solution to this equation is illustrated in figure 3.7. At the optimum x^* the net revenue from the last unit produced in forestry equals the net cost which falls on reindeer husbandry. If forestry were allowed to choose its own harvest rate, it would harvest until marginal net revenue is zero, i.e. x^{F} would be produced. Thus, we see that the private optimum results in a level of forestry activity

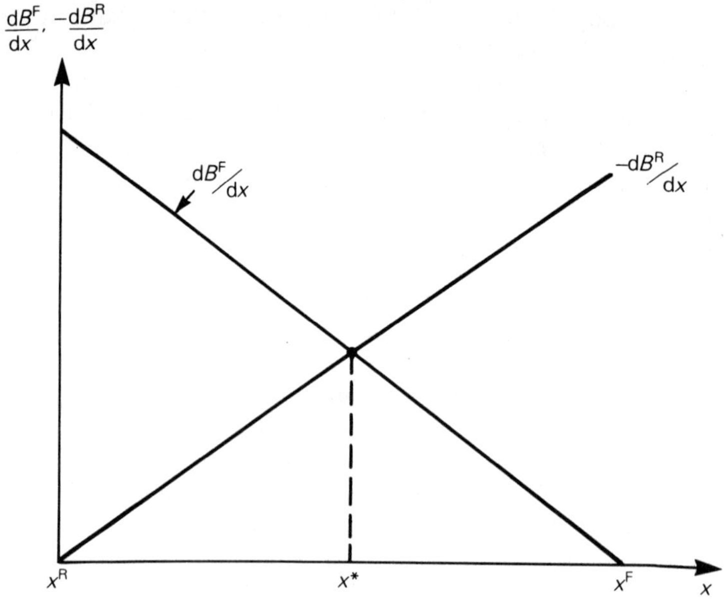

FIGURE 3.7 Illustration of Coase's theorem

which is too large. On the other hand, if reindeer husbandry was allowed to decide the rate of forestry production, forestry would not be allowed at all, i.e. point x^{R} would be chosen.

This land use conflict may be solved by internalizing the externality through letting forestry also conduct reindeer husbandry.[18] The goal functions of society and forestry under such circumstances would coincide. Other possibilities that are available to solve the conflict are direct regulation or the use of taxes in manner discussed in section 3.3.2.

Why should the reindeer husbandry industry quietly allow a level of forestry production which is too large? Both forestry and reindeer husbandry would gain if forestry were compensated to decrease its harvest rate by a sum of money ϵ, where $0 < \mathrm{d}B^{\mathrm{F}}/\mathrm{d}x < \epsilon < -\mathrm{d}B^{\mathrm{R}}/\mathrm{d}x$, so long as the intensity of forestry is larger than x^*. Thus, forestry can be 'bribed' to produce at the socially optimal rate. Moreover, if reindeer husbandry has property rights to the land, forestry can buy the right to use the land by paying $0 < \epsilon - \mathrm{d}B^{\mathrm{R}}/\mathrm{d}x < \mathrm{d}B^{\mathrm{F}}/\mathrm{d}x$, as long as x is smaller than x^*. Hence, property rights do not have any bearing on whether or not the optimal scale of production is

[18] This is to a large extent done in Finland.

attained. However, they do have a bearing on income distribution. These facts are made clear by Coase's (1960) theorem:

Coase's theorem
If costless negotiation is possible, rights are well specified, and redistribution does not affect marginal values, then:

(a) The allocation of resources will be identical, whatever the allocation of legal rights.
(b) The allocation will be efficient, eliminating the problem of externality.

The theorem assumes that transaction costs are zero, and that property rights are well defined. Much of the conflict has been concerned with the property rights of the Laplanders and this might, together with positive transaction costs, contribute to the fact that most people perceive the conflict as being unresolved.

Another important complication in the forestry–reindeer example is the implicit assumption that there are only two parties involved. Assume that negotiations are conducted between one landowner and many reindeer owners. There is an incentive for a reindeer owner to conceal his true willingness to pay in the hope that others will feel sufficiently hurt by forestry and ensure that the 'contract' goes through. In other words, it pays for the individual to be a *free rider*. In economic theory this is often referred to as the 'free rider problem'. If the number of reindeer owners increases, the free rider problem becomes increasingly severe, and the actual scale of forestry approaches the private optimum x^{F}.

Appendix: The Renewable Resource Model and the Price Mechanism over Time

The renewable resource model discussed in section 3.2 obscures price dynamics, since it assumes that the price level is constant over time. Following Berck (1976, 1981) and others, we retain the biomass equation (3.1) in order to obtain a clear insight into the effects of the market and to introduce demand in the simplest possible manner.

The world has one price taking producer who faces a twice continuously differentiable downwards sloping demand function

$$q = D(p) \qquad \frac{\mathrm{d}q}{\mathrm{d}p} = D'(p) < 0 \tag{3.24}$$

where q is the demand for the renewable resource.

We also assume that the producer knows his demand function and uses it to correctly predict the future path of prices.[19] Finally, we assume that the unit cost of harvesting is constant over time, and equal to zero. This is a simple normalization and does not affect the results.

The producer's problem may be written

$$\max_h \int_0^\infty p(t)\, h(t)\, \mathrm{e}^{-rt}\, \mathrm{d}t \tag{3.25}$$

subject to

$$\dot{x} = F(x) - h \qquad h \in [0, \infty)$$

$$x(0) = x_0$$

and the market clearing condition

$$h = D(p)$$

The main difference between the formulation here and the formulation in section 3.2 is that h may now be chosen from an interval with an open right end, and the market always clears. The Hamiltonian is written

$$H = (p\, \mathrm{e}^{-rt} - \lambda)\, h + \lambda F(x) \tag{3.26}$$

The necessary conditions for an optimum are

(a) $\dot{x} = F(x) - h$

(b) $\dot{\lambda} = -\lambda F'(x) \qquad (-\partial H/\partial x = \dot{\lambda})$

(c) h must be chosen to maximize H at every instant.

In section 3.2 the condition that h must be chosen to maximize the Hamiltonian at each moment caused a bang-bang control strategy. However, since h can be set equal to zero or arbitrarily large, it would be suitable to rule out these extremes. The simplest way to do this is to assume that

$$\lim_{p \to \infty} D(p) = 0$$
$$\lim_{p \to 0} D(p) = \infty \tag{3.27}$$

[19] In other words, not only has he 'rational expectations', but also these expectations are fulfilled. For a definition of rational expectation, which is a stochastic concept, see Kantor (1979).

Clearly $h = 0$ will now never be chosen since $h = \epsilon > 0$ would result in a very large revenue, and $h = \infty$ cannot be optimal since it cannot be sold at a positive price.

The only way that h could maximize the Hamiltonian and be different from zero and infinity is when

$$\lambda = p\, e^{-rt} \qquad \left(\frac{\partial H}{\partial h} = 0\right) \tag{3.28}$$

Hence, λ can be interpreted as the present value at time zero of the unit price of the resource at time t. This interpretation is well known from chapter 2. If the present values of the price of the resource were different in two different periods, it would be profitable to reallocate production from the period with the low present value price to the period with the high present value price. By differentiating (3.28) with respect to time,

$$\dot{\lambda} = -rp\, e^{-rt} + \dot{p}\, e^{-rt}$$

and substituting this expression into condition (b) for (3.26) gives

$$\dot{p} = [r - F'(x)]\, p$$

Moreover, by substituting the market clearing condition $h = D(p)$ into condition (a) for (3.26) one obtains

$$\dot{x} = F(x) - D(p)$$

and the necessary conditions for an optimal path can be summed up in the following differential equations:

$$\begin{aligned} &\dot{x} = F(x) - D(p) \\ &\dot{p} = [r - F'(x)]\, p \\ &x(0) = x_0 \\ &\lim_{t \to \infty} p\, e^{-rt} x = 0 \qquad \text{(transversality or terminal condition)} \end{aligned} \tag{3.29}$$

This constitutes two first order differential equations with one initial and one terminal condition, and they define a potentially optimal path $\Phi(t) = [x^*(t), p^*(t)]$. A steady state, i.e. $\dot{x} = \dot{p} = 0$, is defined by[20]

$$\begin{aligned} &r - F'(x) = 0 \\ &F(x) - D(p) = 0 \end{aligned} \tag{3.30}$$

[20] Note that $\lim_{p \to 0} D(p) = \infty$ excludes the possibility of $p = 0$ and, hence, $[r - F'(x)]p = 0$ iff $r - F'(x) = 0$.

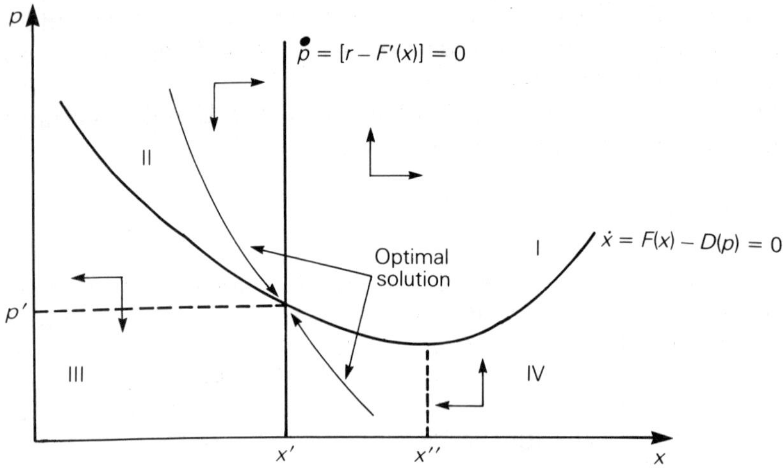

FIGURE 3.8 The optimal solution

Since $r > 0$ and $F''(x) < 0$, the solution to the equation $r = F'(x')$ is unique and identical to the steady state stock in the bang-bang version of the model, provided that the interest rate is the same in the two cases. Moreover, since $D'(p) < 0$ and

$$\lim_{p \to 0} D(p) = \infty$$

and

$$\lim_{p \to \infty} D(p) = 0$$

supply $F(x')$ is equated to demand at a unique price.

The phase arrows in regions I–IV in figure 3.8 indicate how x and p move in these regions, i.e. the signs of $(\dot{x}, \dot{p})$. Suppose $x \neq x'$ and the solution starts in region I. It then holds that $\dot{x} > 0, \dot{p} > 0$. Thus, for a large x, $F'(x) \leqslant 0$ and $\dot{p} > r$, which means that $px\,e^{-rt}$ will increase without bound.[21] Thus, the transversality condition cannot be satisfied. The same holds if the solution starts in III ($\dot{p} < 0$, $\dot{x} < 0$, and $x < 0$ for some $t > t'$). Starting at boundary points ($\dot{x} = 0$ or $\dot{p} = 0$), the solution moves into region I and III and, hence, cannot be optimal.

Thus, given $x(0) = x_0$, $p(0)$ must be chosen in region II or IV, and according to a well known theorem on differential equations

[21] If $\dot{p} > r$, p increases faster than e^{rt} and, hence, $p\,e^{-rt}$ increases without bound.

(see e.g. Pontryagin, 1962) there exists exactly one trajectory that converges to $\dot{x} = \dot{p} = 0$. This solution fulfils the transversality conditions and is therefore a candidate for an optimal solution.[22] In fact, it is the optimal solution.[23]

Proposition 3.3 (Berck, 1976)
The partial equilibrium renewable resource model defined by (3.25) has a unique steady state, and for any $x(0)$ there exists a $p(0)$ such that the model converges to the steady state. The steady state values are implicitly defined by $F'(x') = r$, $F(x') = h'$, and $h' = D(p')$.

If the initial stock is greater (less) than x' the price is 'set' less than p' and p increases (decreases) monotonically at the rate $\dot{p}/p = r - F'(x) > 0$ (<0) while x falls (increases) monotonically.

The word 'set' is in quotation marks because an alternative view is to say that $h(0)$ is chosen with the result that price ends up in these regions. It is established by the bang-bang control model that the optimal steady state stock is a decreasing function of the interest rate. Moreover, since the steady state stock is less than the maximum sustainable yield stock, a decrease of the stock means a lower supply (yield), and a lower supply in steady state means a higher price. (It can also be shown that the optimal trajectory of the shifted system, i.e. the system with $r' = r + \partial r$, will lie below the trajectory of the original system.)

From proposition 3.3 it also follows that price will start below the equilibrium price and rise towards it, and vice versa (sign ∂x = sign $-\partial p$), if the initial stock is larger than the steady state stock.

The common sense explanation for this should be familiar by now to the reader. The price system must create a capital gain to the resource owner to induce him to hold the resource. This is also true for timber management. The value of virgin timber must rise fast enough to offset both the interest rate that the forest owner could earn on the proceeds from cutting down the stand, and the opportunities for new growth.

The Steady State Social Optimum when the Stock is Values

The renewable resource model may be extended in many directions. We will resist all such temptations but one.[24] In natural resource

[22] The steady state is in a strict mathematical sense a so-called saddle point.
[23] For a discussion of sufficient conditions for optimum see e.g. Sydsaeter (1981).
[24] This analysis relies heavily on Berck (1976), (1981). See also chapters 10 and 11.

economics it is often the case that the stock as such has a value to society. Forests are for example used for recreational purposes and a living stock of blue whales has a value in itself to most people. The latter type of value is related to what is known as *option value*.[25] This constitutes no particular problem in a perfect one consumer economy where the flow demand for the resource is generated by a consumer who also values the stock of the resource. Hence, this case may be used to investigate how the social optimum is changed when the externality is taken care of in the proper manner. Real problems emerge when the flow demand of the resource is generated by agents other than those who value the stock. We will soon see that this will, not surprisingly, result in too low a level of stock.

We now extend the above analysis by introducing a utility function which has two arguments. One argument is the flow consumption of the resource, i.e. h, and the other is the stock of the resource x. A forest, like most other natural resources, has the public good attribute of recreation which gives a positive external effect in a general equilibrium setting. We model this by saying that the larger the stock x *ceteris paribus* the higher the level of social utility. To avoid distributional problems, we introduce the assumption that there is only one consumer.

The utility function can be written

$$U = U(h, x) \tag{3.31}$$

and it is assumed to have the proper concavity and differentiability properties to yield a global interior optimum.[26] Society's optimization problem can now be formulated as

$$\max_h \int_0^\infty U(h, x)\, \mathrm{e}^{-rt}\, \mathrm{d}t \tag{3.32}$$

$$\dot{x} = F(x) - h \qquad h \in (0, \infty)$$

$$x(0) = x_0$$

The Hamiltonian is

$$H = \mathrm{e}^{-rt} U(h, x) + \lambda[F(x) - h] \tag{3.33}$$

and the necessary conditions for an optimal path are

[25] For a discussion of different concepts of option value see Hanemann (1984). The concept of option value was originally introduced in economics by Weisbrod (1964).

[26] The function is strictly concave, increasing at each stage of the argument, and twice continuously differentiable

$$\text{(a)}\quad U_h(h,x)\,\mathrm{e}^{-rt} - \lambda = 0 \qquad \left(\frac{\partial H}{\partial h} = 0\right)$$

$$\text{(b)}\quad \dot{\lambda} = -\lambda F'(x) - \mathrm{e}^{-rt} U_x(h,x) \qquad \left(\frac{-\partial H}{\partial x} = \dot{\lambda}\right) \tag{3.34}$$

$$\text{(c)}\quad \dot{x} = F(x) - h$$

$$\text{(d)}\quad \lim_{t\to\infty} \lambda x = 0$$

where $U_h = \partial U/\partial h$, $U_x = \partial U/\partial x$.

If we use the transformation[27] $\lambda = \mathrm{e}^{-rt}p$, we obtain from (3.34a)

$$U_h(h,x) = p$$

and by solving for h we get

$$h = U_h^{-1}(p;x)$$

We can in addition write (3.34b–d) as

$$\text{(a)}\quad \dot{p} = [r - F'(x)]\,p - U_x(h,x)$$

$$\text{(b)}\quad \dot{x} = F(x) - U_h^{-1}(p,x) \tag{3.35}$$

$$\text{(c)}\quad \lim_{t\to\infty} p\,\mathrm{e}^{-rt}x = 0$$

Demand is still independent of time (autonomous), and we can find the locus of points for which $\dot{p} = 0$ by solving (3.35a) where $F(x)$ has been substituted for h. If $U_x[F(x), x]$ decreases in x over the relevant range, it follows that $\dot{p} = 0$ lies to the right of the 'old' $\dot{p} = 0$ locus $[F'(x) = r]$. It is also easy to show that the new $\dot{x} = 0$ curve is raised by including x in the utility function. The new phase diagram can be drawn as in 3.9.

It is clear from the geometry of the problem that the following claim is true:[28]

Proposition 3.4
If the resource stock has a positive value, the socially optimal steady state stock will be higher than it would otherwise have been.

[27] Note that $h = U_h^{-1}(\lambda\,\mathrm{e}^{rt}; x) = D(p; x)$ is the demand function, and that we have to transform the problem so that its solution can be easily compared with the corresponding solution when the stock is not valued.

[28] This result was shown in Berck (1976), (1981). For a similar approach to the optimal rotation problem in forestry see Calish, Fight and Teeguarden (1978), and the analysis in chapter 11.

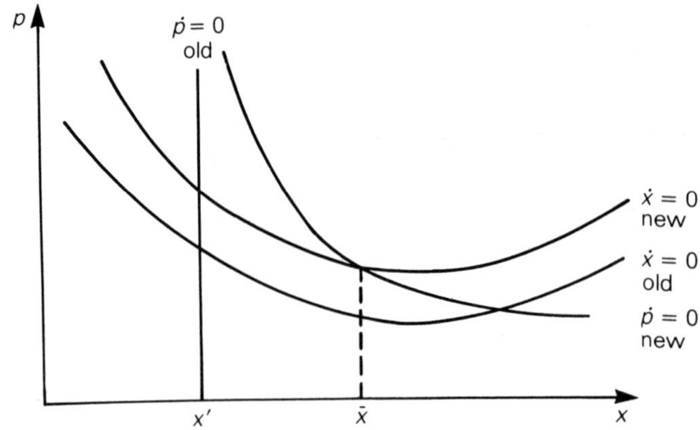

FIGURE 3.9 The steady state with the stock valued

It is not possible to say anything about how the steady state stock of the multiple use system is related to the maximum sustainable yield stock. It could be higher or lower, and the same holds true for the consumption flow in the new steady state.

When the steady state stock coincides with the maximum sustainable yield stock,

$$r\bar{p} = U_x(\bar{h}, \bar{x}) = U_x\,[F(\bar{x}), \bar{x}]$$

Moreover, the price in the steady state is determined from the market clearing condition

$$\bar{h} = U_h^{-1}(\bar{p}, \bar{x})$$

However, from a policy point of view it is not sufficient to know the valuation of a resource unit in the steady state in order to conduct an optimal policy. The market clearing price along the entire trajectory towards the steady state needs to be known, given the initial stock condition, in order to conduct an optimal policy. Translated into the forest policy, it is not sufficient to specify a steady state age distribution; one must in principle specify how to get to the steady state.

On the other hand, given the specification of the one consumer model, with demand derived under the explicit consideration that the stock has a value, the perfect market may be used to generate the optimal path from the society's point of view. This is not possible if flow consumers are different from those who attach a positive or negative value to the stock of the resource. The impatient reader is referred to chapter 10.

4

The Optimal Rotation Period

4.1 Introduction

The classical forest economics problem is the *rotation problem* – that is, the selection of the optimal time to cut an even-aged forest.[1] The profit maximizing solution to this problem, assuming constant price expectations, has in principle been known since the 1849 work of Martin Faustmann. However, as Samuelson (1976) points out, many well known economists have, since Faustmann, published incorrect solutions. The main source of most errors is the neglect of land rent. The purpose of this chapter is to study different aspects of the classical rotation problem, such as how the rotation period is affected by changes in the parameters that determine the solution (prices, wages and interest rates).

Foresters are assumed to maximize the forest land's present discounted value. In section 4.2 we solve the rotation problem under perfectly competitive market conditions. Section 4.3 contains a comparative statics analysis of how the rotation period is affected by changes in the price of timber, regeneration costs, and the internal rate. In section 4.4 the more general problem of assessing the value of forest land is discussed. It is shown that the properties of the present value function can be used to determine the change in the value of forest land that follows from an increased price of timber and decreased planting costs. The subsequent section contains a formal proof that the Faustmann-Pressler-Ohlin (FPO) rotation period is socially optimal. As a byproduct, the optimal consumption pattern over time emerges. The insights gained from this exercise will be used in chapter 7, where we look at some aspects of the behaviour of the self-employed forest farmer. Finally, rotation periods which result from different kinds of cutting criteria – like maximum sustained yield – are compared with the FPO rotation periods in section 4.6.

[1] *Even-aged* refers to timber stands composed of trees of only one age group.

4.2 The Faustmann-Pressler-Ohlin Theorem

In this section we review the economic principles of forestry which are applicable when forestry is treated simply as a source of wood saleable in competitive markets. This means that there are no externalities like environmental problems and/or multiple use considerations.

The so-called rotation problem seems to be very difficult. Many mistakes have been made in the forestry literature and in textbooks of pure economics. Some of the greatest economists have solved the problem incorrectly.[2] The remarkable thing, under these circumstances, is that the problem was formulated correctly in 1849 by Martin Faustmann.[3] The problem was completely solved 11 years later by the German, Max Robert Pressler,[4] who introduced the concept of *das Weiserprozent* (later). It was also solved by the Swedish Nobel prizewinner, Bertil Ohlin, when he was 18 years of age.[5]

A well-defined solution to the rotation problem demands a well-defined problem. We begin with a heroic set of assumptions that yield a simple steady state solution to the rotation problem. Our assumptions are explicit and/or implicit in the works of Faustmann, Ohlin and Pressler. The assumptions are:

Assumption 4.1
The capital market is perfect. One can lend and borrow any amount at the prevailing interest rate. The interest rate is known with certainty over all future periods.

This is clearly a heroic assumption since future interest rates are not known with certainty. In addition, the forest owner cannot borrow and lend unlimited amounts at one interest rate. As we saw in

[2] For an excellent survey of the mistakes, consult Samuelson (1976). See also Löfgren (1983).
[3] Faustmann was born in 1822 in Giessen. He graduated from college as a master of forestry in 1845. In the following four years he passed the examinations necessary for forestry service. In 1857 he became a conservator for the Grand Duchy of Hessen and he remained in this position till his death in 1876.
[4] See Pressler (1860). Pressler was born in 1815 in Dresden. He graduated from the Institute of Technology in Dresden in 1835. At the age of only 25, he became Professor of Agricultural and Forest Techniques at the Academy of Agriculture and Forestry in Tharandt. He remained there till 1883, when he retired. He died in 1886.
[5] Ohlin (1921). Ohlin, who was born in 1898, is well known for his contributions to international trade theory – the so-called *Heckscher-Ohlin theorem*, which states that a country will export the goods that are intensive in its abundant production factor. He was awarded the Nobel prize for economics in 1978, one year before his death.

chapter 1, an imperfect capital market means that the forest owners' preferences contribute to optimal decisions in forestry.

Assumption 4.2
All future timber prices and the future prices of all inputs are constant and known.

Assumption 4.3
Future timber yields, which are functions of once certain expenditure inputs (planting, sprayings, thinning etc.), are known.

Assumption 4.3 is also restrictive. Stochastic climate factors – such as lightning, forest fires and disease – must in the real world affect timber yields.[6] Changes in growing technologies and utilization rates also have important effects on timber yields.

Assumption 4.4
Forest land can be bought, sold and rented in a perfect market.

In addition, we assume that there are no restrictions on cutting capacity or on the composition of the entire forest; this allows us to work at the stand level.

Given assumptions 4.1–4.4, the solution of the rotation problem at the stand level will be a steady state solution, i.e. each cycle will be of equal length.

Let

$$Q_{t+T} = f(T, M)$$

be the growth or production function, which is given by technological and biological considerations. Q_{t+T} is the output when the forest is cut at time T, and M is the area of land that is used throughout the complete cycle. For convenience, we assume that $M = 1$ and that we are dealing with an even-aged stand. The growth function is assumed to have the following properties:

(a) $f \in C^2$

(b) $f'(T) > 0 \qquad 0 < T < a,\ f'(a) = 0$

(c) $f''(T) < 0$

[6] For a stochastic version of the rotation problem see Brock, Rothschild and Stiglitz (1979). See also chapter 12.

The growth function is continuously differentiable twice and increasing in T in the open interval $(0, a)$. At $T = a$ the forest land contains the maximum amount of standing timber, and the amount of timber may well decrease for $T > a$, i.e. $f'(a + \epsilon) < 0$. The function f may be thought of as containing labour as an input, but the quantity of labour needed to recycle one unit of land is constant over time. Technologies of this type, used by Jevons (1871), Wicksell (1901) and Lutz and Lutz (1951) are often referred to as *point-input/point-output technologies*.

The assumption $f''(T) < 0$ means that the growth rate $f'(T)$ decreases over time. This is a simplification. The growth function of timber has a section of increasing returns $f''(T) > 0$, followed by a section of decreasing returns $f''(T) < 0$, as in figure 4.1. This function expresses the volume per unit land area as a function of time. At around $T = 15$ years there is a notch in the curve since at that time the forest owner undertakes a clearing. At $T = 45$ there is another notch in the function. At this juncture, the first thinning is

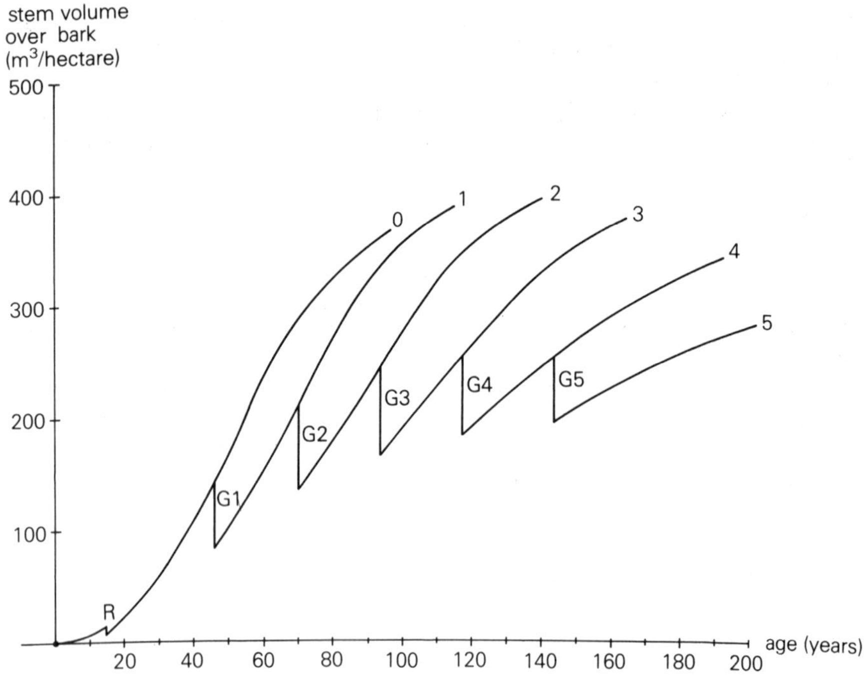

FIGURE 4.1 The growth function for *Pinus silvestris*

undertaken. The notches denoted G_2, G_3, G_4 and G_5 represent later thinnings. The growth function where no thinning is undertaken is marked with a zero.

Expressing the change in volume over time ignores all the biological and climatological factors which determine growth. A function incorporating these factors is

$$Q_{t+T} = f(T; \beta)$$

where β is a parameter vector containing 'biological variables' which describe the properties of the stand and the environment.

An example of what β might contain is given in the growth function for Norway spruce estimated by Eriksson (1976):

$$\begin{aligned} Q_{t+T} &= T^{b_1} f(b; a, K) \\ &= aT^{b_1} G_1^{b_2} (th + 0.01)^{b_3} [(D_3 D_1^{-1} + 0.1)(100 N_3 N_1^{-1} \\ &\qquad + 0.01)]^{0.062} H_D^{b_4} V_G^{b_5} \end{aligned}$$

where the variables have the following meanings:

Q_{t+T}	annual accumulated volume increment under bark per hectare (m^3)
G_1	basal area under bark per hectare before thinning (m^2)
th	thinning percentage of basal area under bark
D_3	diameter corresponding to mean basal area under bark of the thinned tree (cm)
D_1	diameter corresponding to mean basal area of the stand before thinning (cm)
N_3	number of thinned trees per hectare
N_1	number of trees per hectare in the stand before thinning
H_D	dominant height (dm)
T	age at breast height of the 100 largest trees per hectare (years)
V_G	group structure index
a	a parameter
$b = (b_1, \ldots, b_5)$	a parameter vector where the parameters vary with the site index
K	$[G_1, D_1, D_3, N_1, N_3, H_D, V_G, th]$ a vector describing the 'biological status' of the stand

Since some of the included variables vary with T, it is in fact not entirely reasonable to view the growth as a function of T with a and the b_i as given constants.[7] The function illustrates the complexities involved in yield estimates in practical forestry.[8] However, since this chapter is concerned with the qualitative aspects of forestry eonomics, the above parameters $[a, b, K]$ will be excluded from the following analysis.

The following analysis contains many simplifications. Firstly, neither thinnings nor clearings are options for the forest owner. This means that clearing and thinning times are exogenously determined and that one therefore can disregard any technical discontinuity problems which are created by these activities. To treat them as endogenous variables is fairly difficult and provides few new economic insights.[9] Secondly, we assume that a cubic metre of wood has the same value, independent of the age of the tree.

The present value of profit over one rotation period can be written

$$\pi(T) = pf(T)\,\mathrm{e}^{-rT} - wl \tag{4.1}$$

where w is the wage rate, p is the net price of a cubic metre of wood (net of cutting costs), l is the input of labour, and r is the interest rate.

The term wl may be thought of as planting costs, and $pf(T)\,\mathrm{e}^{-rT}$ is the present value of the cut.

If (4.1) is maximized with respect to T, the first order condition for an interior optimum can be written

$$f'(T_{\mathrm{H}})/f(T_{\mathrm{H}}) = r \tag{4.2}$$

which means that *a forest stand shall be cut when its relative value growth equals the interest rate*.[10] This formula also appeared in sections 2.3 and 3.2. T_{H} is the one cycle rotation period.

This rule seems intuitively reasonable. It is obviously not profitable to maintain the forest when its relative value growth is lower than the interest rate. It would then be profitable to cut the forest and deposit the timber proceeds in a bank account. It also seems reasonable to maintain a forest stand whenever its relative value growth exceeds the interest rate. This reasoning, however, is incorrect

[7] $Q_{t+T} = f(\bar{b}, \bar{a})\,T^{b_0} = AT^{b_0}$ is probably unreasonable as A is not a constant but a function of T. $H_{\mathrm{D}}^{b_4}$ is clearly a function of T.

[8] For a discussion of recent developments, see Hägglund (1981).

[9] For an explicit treatment of the rotation problem, when thinning is treated as an endogenous variable, see Näslund (1969). See also Clark (1976).

[10] Note that we use the terminology *value growth* although no prices are present in the formula. In a more general setting where the unit value of wood depends on age, prices in the denominator and numerator of (4.2) will no longer cancel out.

since the forest must be cut before the land can be used again. Each year the cutting is postponed means that all future proceeds will be realized one year later. This implies a cost to the owner of the forest, and this cost must be deducted from future sales proceeds before the relative value growth is compared with the interest rate.

Another way of viewing the incorrectness of (4.2) is to realize that the forest owner has the option of cutting the forest and letting out the land to another forester. The extra cost from lengthening a rotation cycle by one year is then equal to the value of the rent that would have been paid during the year. This is another way of expressing the reduction of the relative value growth that must occur before it is compared with the prevailing interest rate. We will see later (section 4.5) that the yearly rent in a perfect capital market equals the interest on the value of the land.

We now introduce the correct solution to the rotation problem. The present value of the forest land is the sum of the infinite series of all future revenues:[11]

$$\begin{aligned} V(T) &= [pf(T)\,\mathrm{e}^{-rT} - wl]\,[1 + \mathrm{e}^{-rT} + (\mathrm{e}^{-rT})^2 + \ldots] \\ &= [pf(T)\,\mathrm{e}^{-rT} - wl]\,(1 - \mathrm{e}^{-rT})^{-1} \qquad (4.3) \end{aligned}$$

Formula (4.3) assumes that it is optimal to choose all rotations of equal length. This is true, since the optimization problems at time zero (newly seeded land) and after the first rotation has been completed ($T = T_1$) look exactly the same, owing to the infinite planning horizon and stationary conditions. If the first rotation was optimally chosen in the first place, it is obvious that the second rotation chosen should be equal to the first, i.e. $T_1 = T_2$. The argument can be repeated for the third rotation and so on.

The next step, not taken by Faustmann, is to choose T in such a way that V is maximized, although there are indications in his paper that he had the right idea. The first solution is probably due to Pressler (1860) and an early solution to the maximization problem was also given by Beril Ohlin in 1917.[12]

The first order condition for a maximum forest present value is obtained by differentiating (4.3) with respect to T, and setting the

[11] Formula (4.3) is known to foresters as *Faustmann's formula* and was derived by the German Martin Faustmann in 1849. The purpose of Faustmann's paper was to correct an earlier attempt by E. F. von Gehren to use a present value calculation to determine the value of forest land.

[12] See Löfgren (1983). Ohlin presented his solution at a seminar at the Stockholm School of Economics led by Heckscher. In the theory of investment, the analogous problem is known as the chain of replacement problem. See e.g. Masse (1962).

resulting derivative equal to zero:

$$\frac{\mathrm{d}V(T)}{\mathrm{d}T} = [pf'(T)\,\mathrm{e}^{-rT} - rpf(T)\,\mathrm{e}^{-rT} - rV(T)\,\mathrm{e}^{-rT}]\,(1-\mathrm{e}^{-rT})^{-1} = 0$$

For an interior solution ($T > 0$), this condition can be simplified to read

$$pf'(T^0) = rpf(T^0) + rV(T^0) \tag{4.4}$$

yielding

Theorem 4.1 (Faustmann-Pressler-Ohlin)
A forest stand shall be harvested when the rate of change of its value with respect to time is equal to the interest on the value of the stand plus interest on the value of the forest land.[13]

4.3 Comparative Statics

The optimal rotation period T^0 is a function of the parameters of the problem p, w, r. Hence (see also Heaps, 1981)

$$T^0 = T^0(p, w, r) \tag{4.5}$$

The one cycle (von Thünen) rotation period from (4.2) can also be written as a function of the parameters.[14] However, from the first order condition (4.2) it is clear that the one cycle rotation period does not depend on p and w. Hence

$$T_{\mathrm{H}} = T_{\mathrm{H}}(r) \tag{4.6}$$

The second order maximization conditions require that the present value functions in both cases are strictly concave at the interior optimum. For this to be the case, it is sufficient to assume that the biotechnology is strictly concave, i.e.

$$f''(T) < 0 \tag{4.7}$$

This property of biotechnology will be used in the comparative statics to follow.

[13] Pressler's *Weiserprozent* is defined as $i^* = [f'(T)/f(T)] - [r/(\mathrm{e}^{rT}-1)]$, where $f'(T)$ is the first derivative of $f(T)$ with respect to T. It is thus equal to the rate of growth of the forest stand corrected for interest on the land value divided by the current value of the stand. The relationship with (4.4) should be clear. The correct cutting rule is to cut the forest stand when $i^* \leqslant r$ and save the stand as long as $i^* > r$.

[14] The famous German economist J. H. von Thünen had addressed the rotation problem a few years before Faustmann, and he ended up with the suggestion explicit in (4.2) (von Thünen, 1863).

We begin with the relationship between the true competitive solution and the one cycle result. For an interior solution, i.e. $T^0 > 0$, the present value of the forest land must be positive,[15] i.e. $V > 0$. It follows then from a comparison of the two first order conditions (4.2) and (4.4), and the strict concavity of the production function, that $T^0 < T_H$. This result was also implicit in the earlier qualitative discussion, which established that the rule that the forest stand shall be cut down when its relative value growth coincides with the interest rate is incorrect.

We are now ready to analyse the comparative statics properties of the behaviour equations (4.5) and (4.6). If we begin with (4.6) – the 'incorrect' solution – it is easily shown that

$$\frac{\mathrm{d}T_H}{\mathrm{d}r} < 0$$

i.e. an increase in the interest rate shortens the one cycle rotation period. To see this, totally differentiate the equilibrium condition (4.2):

$$[f''(T_H) - rf'(T_H)]\,\mathrm{d}T_H = f(T_H)\,\mathrm{d}r$$

Hence

$$\frac{\mathrm{d}T_H}{\mathrm{d}r} = \frac{f(T_H)}{f''(T_H) - rf'(T_H)} < 0 \quad (f'' < 0,\ f' > 0,\ f > 0)$$

It is also clear that, as long as the present value remains positive, no taxes (profit and/or sales tax), price or wage changes can affect the rotation period.

We now turn to the correct solution. To investigate the properties of (4.5), we differentiate the equilibrium condition (4.4) totally[16]

$$[pf''(T^0) - rpf'(T^0)]\,\mathrm{d}T$$

$$= \left(rf(T^0) - f'(T^0) + \frac{rf(T^0)\,\mathrm{e}^{-rT^0}}{1 - \mathrm{e}^{-rT^0}} \right) \mathrm{d}p \qquad (A)$$

$$- \left(\frac{rl}{1 - \mathrm{e}^{-rT^0}} \right) \mathrm{d}w \qquad (B) \quad (4.8)$$

$$+ \left(pf(T^0) + \frac{\mathrm{d}}{\mathrm{d}r}\,[rV(T^0; r)] \right) \mathrm{d}r \qquad (C)$$

[15] Otherwise the forest owner would abstain from planting.

[16] Note that $\partial V(T^0)/\partial T = 0$ as the value of land is maximized at T^0.

The left hand side of (4.8) is negative since $f''(T) < 0$ and $f'(T) > 0$. Term B on the right hand side of (4.8) is negative. Term A can be shown to be positive by making use of (4.4). Hence

$$\frac{\partial T^0}{\partial p} < 0 \qquad (\mathrm{d}w = \mathrm{d}r = 0)$$

$$\frac{\partial T^0}{\partial w} > 0 \qquad (\mathrm{d}p = \mathrm{d}r = 0)$$

which validates a common sense interpretation that an increase in the future price of timber[17] *ceteris paribus* increases the value of the forest land which, in turn, increases the interest cost of lengthening the rotation period. To compensate for this interest cost, the forest owner must shorten the rotation period in order to obtain a larger value growth.

An increase in the wage rate (planting costs) decreases the value of the forest land *ceteris paribus* and decreases the interest costs on the value of land. (This result presupposes that labour use is fixed.) As a result, the rotation period lengthens and the value growth of the forest declines.

If wages and prices are scaled by a constant $\lambda > 0$, thereby keeping the real wage wp^{-1} constant, one has

$$\mathrm{d}w = \frac{w}{p}\,\mathrm{d}p$$

Substituting for dw in (4.8) yields

$$[pf''(T^0) - rpf'(T^0)]\,\mathrm{d}T = (A + Bw)\,p^{-1}\,\mathrm{d}p$$

and $A + Bw = 0$ from the first order condition (4.4). Hence

$$T^0(p, w, r) = T^0(\lambda p, \lambda w, r)$$

for $\lambda > 0$. Letting $\lambda = p^{-1}$, (4.5) can be written

$$T^0 = T^0\left(\frac{w}{p}, r\right) \tag{4.5a}$$

Hence the FPO rotation period depends only on relative prices, not on the absolute levels of p and w.

[17] A possible interpretation is that a forest owner has unitary elastic price expectations, i.e. if price (value of *stumpage* or standing timber) changes in the next period by 1 per cent all future prices are expected to change by 1 per cent.

To determine whether an increase in the interest rate increases or decreases T^0, it is necessary to see how the interest cost on the value of land is affected by an increase in the interest rate. We know that

$$\frac{\mathrm{d}}{\mathrm{d}r}[rV(T^0;r)] = V(T^0;r) + r\frac{\partial V(T^0,r)}{\partial r} \gtrless 0 \tag{4.9}$$

The sign of (4.9) is indeterminate since an increase in r increases (4.9) by $V(T^0; r)$. At the same time it decreases the present value of the forest land by $r(\partial V/\partial r)$.

Rewrite (4.9) to obtain

$$\frac{\mathrm{d}}{\mathrm{d}r}[rV(T^0;r)] = V(T^0;r) - \frac{rT^0 pf(T^0)\,\mathrm{e}^{-rT^0}}{1-\mathrm{e}^{-rT^0}} - \frac{rT^0\,\mathrm{e}^{-rT^0}\,V(T^0;r)}{1-\mathrm{e}^{-rT^0}}$$

Substitute this expression into C in (4.8) to get

$$C = [V(T^0;r) + pf(T^0)]\left(1 - \frac{rT^0}{\mathrm{e}^{rT^0}-1}\right)$$

$V(T^0;r) > 0$ means that $C > 0$ if and only if

$$c = \left(1 - \frac{rT^0}{\mathrm{e}^{rT^0}-1}\right) > 0$$

and this is true for all $rT^0 > 0$. For rT^0 close to zero,

$$\lim_{rT^0 \to 0} c = 0$$

We conclude that $C > 0$ for $V(T^0; r) > 0$ and this means that $\partial T^0/\partial r < 0$. In sum:

Proposition 4.1
The FPO optimal rotation period T^0 is an almost everywhere differentiable function of the parameters p, w, r:

$$T^0 = T^0(p, w, r)$$

The partial derivatives are

(a) $\dfrac{\partial T^0}{\partial p} < 0$

(b) $\dfrac{\partial T^0}{\partial w} > 0$

(c) $\dfrac{\partial T^0}{\partial r} < 0$

(d) $T^0(\lambda p, \lambda w, r) = T^0(p, w, r)$ (homogeneity of degree zero)

Since the optimal rotation period is a function of p, w and r, the land value function can be written as

$$\max_T V(T; p, w, r) = V[T^0(p, w, r); p, w, r]$$

To analyse how land values are affected by changes in the parameters, we must consider two effects. An increase in p has a direct effect on $V(\cdot)$ because the value of timber increases during the rotation period. Increasing p has an indirect effect on $V(\cdot)$ since the optimal rotation period is shortened when p changes. However, we now show that the indirect effect can be neglected:[18]

Proposition 4.2
Given the assumptions behind proposition 4.1, then:

(a) $$\frac{\mathrm{d}}{\mathrm{d}p}\{V[T^0(p, w, r); p, w, r]\} = \frac{\partial}{\partial p} V[T^0(p, w, r); p, w, r]$$

(b) $$\frac{\mathrm{d}}{\mathrm{d}w}\{V[T^0(p, w, r); p, w, r]\} = \frac{\partial}{\partial w} V[T^0(p, w, r); p, w, r]$$

(c) $$\frac{\mathrm{d}}{\mathrm{d}r}\{V[T^0(p, w, r); p, w, r]\} = \frac{\partial}{\partial r} V[T^0(p, w, r); p, w, r]$$

Proof
The function $V(\ )$ is differentiable. Hence

$$\frac{\partial V}{\partial T^0}\frac{\partial T^0}{\partial p} + \frac{\partial V}{\partial p} = \frac{\partial V}{\partial p}$$

since $\partial V/\partial T^0 = 0$ (T^0 is an interior maximum). This proves the first claim, and provides an excellent clue on how to prove the remaining claims.

The interpretation of (a) in proposition 4.2 is that the *total effect* on present value of a once and for all increase of timber prices[19] is equal

[18] In a generalized form this mathematical result has many applications in economics. It was probably Hotelling (1932) who first used this property of a function which is maximized. The property was rediscovered by Samuelson (1947) and used by Shephard (1953) to derive the so-called 'Shephard's lemma'. The general mathematical result, which implies both Hotelling's result and Shephard's lemma, is called the envelope theorem. See Varian (1978).

[19] Note that price is changed for all future periods.

to the *partial* derivative of present value with respect to the price:

$$\frac{\mathrm{d}V}{\mathrm{d}p} = \frac{\partial V}{\partial p} = \frac{f(T^0)}{\mathrm{e}^{rT^0} - 1} = \frac{Q_{t+T^0}}{\mathrm{e}^{rT^0} - 1} > 0 \tag{4.10}$$

which is the sum of the supplies in cubic metres,[20] discounted by the interest rate r. Proposition 4.2 has important implications for the assessment of forest land values. For reasonably small changes in the parameters, there is no need to assess the effect on land values from the changed rotation period.

4.4 The Assessment of the Value of Forest Land

In the previous analysis we disregarded the value of standing timber by implicitly assuming that the valuation concerns only newly seeded land. As will be seen later, it is fairly easy to generalize the analysis and take explicit account of the value of the standing timber. Given the growth function $f(T)$, the value of standing timber of age b can be written $pf(b)$. This stand should optimally be cut at $T \geqslant b$, and the present value at date b of this revenue is $pf(T)\,\mathrm{e}^{-r(T-b)}$ (see figure 4.2).

Let us further assume that this stand is immediately regenerated, i.e. the cut is a joint product which includes regeneration. Hence we can ignore explicit regeneration costs. If the same biotechnology is used during every period, the net revenue series is

$$pf(T')\,\mathrm{e}^{-rT'},\; pf(T')\,\mathrm{e}^{-r2T'}, \ldots$$

The present value at the time for regeneration of this series is

$$V = \frac{pf(T')\,\mathrm{e}^{-rT'}}{1 - \mathrm{e}^{-rT'}} = \frac{pf(T')}{\mathrm{e}^{rT'} - 1} \tag{4.11}$$

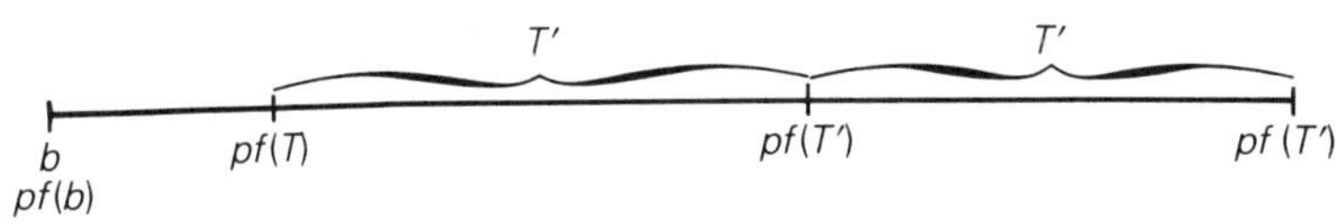

FIGURE 4.2 The revenues over time

[20] The term *supplies* refers to the cut in cubic metres at $T^0, 2T^0, 3T^0, \ldots$, which is an infinite series.

According to Bellman's (1957) principle of optimality, every part of the optimal path must be optimal. We can therefore determine

$$\max_{T'} V(T')$$

independently of how the cutting problem of the standing timber is solved, i.e. the optimal rotation period $T' = T^0$ is determined in the same manner as in section 4.2. Denote the optimal value of seeded land by $V(T^0)$. In order to determine when it is optimal to cut the standing timber, we maximize the present value of land plus the value of the standing timber:

$$\max_{T} PV = [R(T) + V(T^0)]\, \mathrm{e}^{-r(T-b)} \tag{4.12}$$

where

$$R(T) = pf(T)$$

The first order condition for a maximum can be written

$$pf'(T) = rpf(T) + rV(T^0)$$

which, considering (4.4), is fulfilled for $T = T^0$. The optimal value of the forest land is

$$PV^0 = \left[pf(T^0) + \frac{pf(T^0)}{\mathrm{e}^{rT^0} - 1} \right] \mathrm{e}^{-r(T^0 - b)} = \frac{\mathrm{e}^{rb} pf(T^0)}{\mathrm{e}^{rT^0} - 1} = \mathrm{e}^{rb}\, V(T^0) \tag{4.13}$$

For seeded land $b = 0$, and

$$PV^0 = V(T^0)$$

For a mature stand $b = T^0$, and

$$PV^0 = V(T^0)\, \mathrm{e}^{rT^0} = pf(T^0) + V(T^0)$$

Hence, when the land contains standing timber, the land value is obtained by scaling up the value of bare land by a factor e^{rb}, where b is the age of the standing timber.[21]

Now assume that prices increase by 10 per cent and that this is known by all economic agents. Will this price increase cause the value of forest land to increase by more or less than 10 per cent? To answer this question we have to evaluate whether

$$\epsilon_p = \frac{\mathrm{d}PV/PV}{\mathrm{d}p/p} = \frac{\mathrm{d}PV}{\mathrm{d}p} \frac{p}{PV} \gtreqless 1$$

[21] This was realized by Faustmann (1849), p. 446.

where ϵ_p is the price elasticity of the value of the forest land, $\mathrm{d}PV/PV$ is the relative increase in value, and $\mathrm{d}p/p$ is the relative increase in the price level. From proposition 4.2, (4.10) and (4.13),

$$p\,\frac{\mathrm{d}PV}{\mathrm{d}p} = p\,\frac{\partial PV}{\partial p} = \frac{pf(T^0)\,\mathrm{e}^{rb}}{\mathrm{e}^{rT^0} - 1} = PV$$

Hence, a 10 per cent increase in the price level will increase land value by 10 per cent. Note, however, that with regeneration costs positive, $\epsilon_p > 1$, a 1 per cent increase in the price level will increase land values by more than 1 per cent. Similarly a 1 per cent decrease in regeneration costs will increase land values by more than 1 per cent The reader is invited to compute the interest elasticity of the value of forest land, making use of the envelope property $\mathrm{d}PV/\mathrm{d}r = \partial PV/\partial r$.

4.5 The Social Optimality of the FPO Rotation Period

In this section we provide normative reasons for our extensive concern with the FPO rotation period. We show, within a partial equilibrium setting, that the FPO rotation period is socially optimal given assumptions 4.1–4.4. The method used to derive the explicit solution to the optimization problem is optimal control theory. Although the mathematics is somewhat complex, the intuition behind the optimal solution is quite simple.

Consider a utility function

$$U = U[c(t)] \quad U'[c(t)] = \frac{\partial U[c(t)]}{\partial c} > 0 \text{ all } c \tag{4.14}$$

This utility function may be the utility function of the forest owner or, better, the utility function of a sole consumer in a one consumer society. The complications caused by additional agents and goods are mainly of a distributional character and do not change the core of the results. $c(t)$ denotes real consumption at time t, and the particular form of the utility function means that utility at time t is only a function of consumption at time t.[22] $U'[c(t)] > 0$ means that the marginal utility of the last unit consumed is positive; more is always better than less.

[22] One could easily imagine that utility at time t could depend on consumption both at t and at some earlier moment. This complication is usually ducked but, even explicitly considered, it would – in most instances – not affect the social optimality of the FPO rotation period.

The household maximizes the present value of the sum (the integral) of total utility over time:

$$\max_{c(t)} \int_0^\infty U[c(t)]\, e^{-rt}\, dt \tag{4.15}$$

Since we assume perfect capital markets, the *rate of time preference*[23] equals the interest rate. Hence, utility is properly discounted by the interest rate r. The integral (4.15) is maximized subject to the intertemporal budget constraint that the stream of revenues from forestry is greater than or equal to the present value of real consumption:

$$K = \frac{f(T)}{e^{rT} - 1} \geqslant \int_0^\infty c(t)\, e^{-rt}\, dt \tag{4.16}$$

We assume that wood and goods exchange at unit prices, and that regeneration costs are equal to zero. The optimality of the FPO rotation is not violated if these assumptions, excepting perfect capital markets, are relaxed.

K, or capital, is the present value of the net incomes from forestry, given that society has selected a uniform but not necessarily optimal rotation T. K is also the upper bound on the consumption possibilities and, from what we learned in chapter 1, this upper bound should be chosen as large as possible. K is also independent of the consumption pattern over time. We maximize K by choosing T such that

$$K^0(r) = \max_T \left[\frac{f(T)}{e^{rT} - 1}\right] \tag{4.17}$$

which yields the FPO rotation period. Hence the FPO rotation period is socially optimal.

The optimal consumption path is determined by solving the optimization problem in (4.15) and (4.16) using the equation of motion

$$\dot{K}^0 = -c(t)\, e^{-rt} \qquad \dot{K}^0 = \frac{dK^0}{dt} \tag{4.18}$$

where the dot is used to denote derivatives with respect to time.

The Hamiltonian of the optimal control problem is

$$H = e^{-rt}\{\lambda_0(t)\, U[c(t)] - \lambda(t)\, c(t)\} \tag{4.19}$$

[23] Defined as the minimum premium the consumer must receive before he will postpone a dollar's worth of consumption in one period

where $\lambda_0(t)$ and $\lambda(t)$ are costable variables, and $\lambda_0(t)$ is a constant which can be set equal to one since the right hand endpoint is free. An 'interior' solution to this problem must fulfil

$$\frac{\partial H}{\partial c(t)} = U'[c(t)] - \lambda(t) = 0$$
$$\frac{\partial H}{\partial K^0} = -\dot{\lambda}(t) \tag{4.20}$$

From (4.20) and the Hamiltonian ($\partial H/\partial K^0 = 0$), it follows that consumption is constant over time:

$$c(t) = U'^{-1}(\lambda) = \bar{c}(\lambda) \quad \mathrm{d}c/\mathrm{d}\lambda < 0^{24} \tag{4.21}$$

This follows because the shadow price of consumption λ is constant over time, as seen in (4.20). Substituting (4.21) into (4.16) yields

$$K^0 = \bar{c}(\lambda) \int_0^\infty \mathrm{e}^{-rt}\,\mathrm{d}t = \bar{c}(\lambda)/r \tag{4.22}$$

and

$$\bar{c}(\lambda) = rK^0$$

Hence, in each 'period' a welfare maximizing society consumes the constant return rK^0 (income)[25] from its renewable resource, although this resource is harvested only at times T^0, $2T^0$ etc. In other words, consumption decisions are completely separated from production decisions. In sum:

Proposition 4.3
In the perfect market optimization problem specified by (4.15) and (4.16), the FPO rotation period is socially optimal. Real consumption is constant over time and equals the interest on the land value rK^0 (income).

4.6 The FPO Rotation Period Compared with other Periods

We saw in section 4.2 that the correct rotation period T^0 is shorter than the Thünen-Wicksell rotation period T_{H} which results from the

[24] Follows from the strict concavity of the utility function.
[25] The stringent definition of income is interest on the capital.

maximization of the present value over one rotation period.[26] In this section we introduce three additional often proposed, but 'incorrect', investment criteria and compare the resulting rotation periods with the true competitive solution T^0.

We begin with an investment criterion that has been proposed by Boulding (1935). The internal rate of return was defined in chapter 1. Boulding suggests that the optimal rotation period is the T that maximizes the internal rate of return. Given T, the internal rate of return is defined as the interest rate that equates the present value of the proceeds from the timber sales at T and the initial planting costs.

Formally, the internal rate of return r_B is implicitly defined by

$$pf(T)\,e^{-r_B T} = wl \tag{4.23}$$

Solving (4.23) for r_B,

$$r_B = T^{-1} \log \frac{pf(T)}{wl} \tag{4.24}$$

and the first order condition for a maximum can be written

$$f'(T_B)/f(T_B) = T_B^{-1} \log \frac{pf(T_B)}{wl} \tag{4.25}$$

This equation is obtained by differentiating (4.24) with respect to T. Comparing (4.2) and (4.25) we see that $r = T_B^{-1} \log [pf(T_B)/wl]$ implies $T_H = T_B$, where T_H is the Thünen-Wicksell optimal 'wine aging' period. Moreover, since $\partial T_H/\mathrm{d}r < 0$, it follows that $r_B > r$ implies $T_B < T_H$.

Among foresters it has often been claimed that the ultimate goal of forest management should be to maximize some kind of maximum sustained yield. Samuelson (1976) has suggested that it is not always clear what one means by the maximization of sustained yield. The interpretation given by most foresters is that one wishes to maximize the gross yield defined by

$$G = f(T)\,T^{-1} \tag{4.26}$$

i.e. one maximizes the average output over the rotation period. In a sense, this is a very natural way to solve the present value maximization problem when the interest rate is zero. If the interest rate is

[26] Wicksell (1901) solves the wine aging problem, for which T_H is a reasonable, not to say correct, rotation period provided that storage capacity is for all practical purposes a non-binding restriction.

zero, the land value is not defined and it would then be sensible to maximize the average revenue over the rotation period. When the price level is given, such behaviour is equivalent to the maximization of $G(T)$ in (4.26). The optimality of this action can be readily shown by proving that the Faustmann-Pressler-Ohlin (FPO) rotation period approaches the rotation period which maximizes gross (net)[27] sustainable yield when the interest rate tends to zero. The gross sustained yield maximum is obtained for the T_G which solves the equation

$$\frac{d}{dT}[f(T)\,T^{-1}] = \frac{f'(T)\,T - f(T)}{T^2} = 0$$

For an interior solution ($T_G > 0$) it must hold that

$$f'(T_G) = f(T_G)\,T_G^{-1} \tag{4.27}$$

Hence the current annual increment $f'(T_G)$ equals the mean annual increment $f(T_G)/T_G$.

Intuition suggests that (4.27) has the following meaning. As long as the current annual increment is larger than the mean annual increment, the mean will increase over time. Similarly, when the current annual increment is less than the mean, the mean will decrease. In other words, if the mean annual increment reaches a maximum, it must be equal to the current annual increment at the maximum. This is illustrated by figure 4.3, which is generated from a growth function with a section of increasing returns, $f''(T) > 0$, followed by a section of decreasing returns, $f''(T) < 0$ (see also figure 4.4).

The maximum sustained gross yield is not defined as the rotation period for which $f'(T) = 0$, i.e. $T = a$. It is clearly unreasonable for a forest owner to wait until every tree is fully grown until harvest time since it is profitable some time before $T = a$ to cut the forest and make the land available for faster growing younger trees.

Given an account of planting costs, a *net* sustainable yield concept - what Gregory (1972) calls *forest rent* - is defined as

$$N = \left[f(T) - \frac{w}{p}\,l\right]T^{-1} \tag{4.28}$$

The first order condition for an interior maximum is

$$f'(T_N) = f(T_N)\,T_N^{-1} - \frac{w}{p}\,l T_N^{-1} \tag{4.29}$$

[27] Gross sustainable yield if planting costs are zero and net sustainable yield if planting costs are positive.

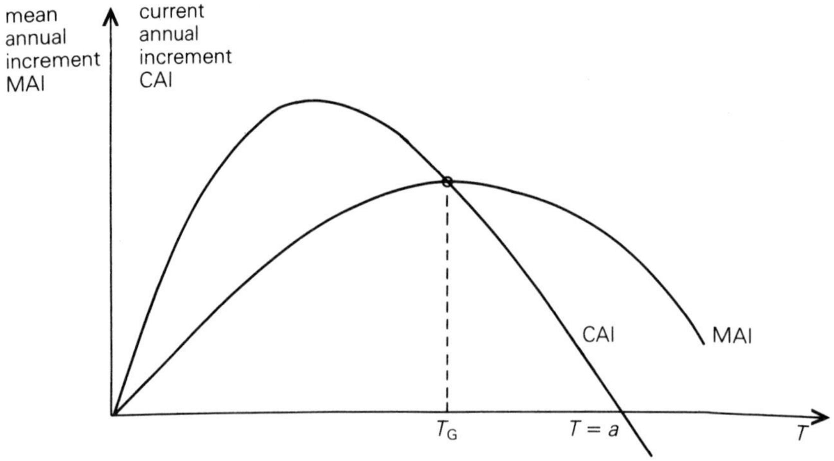

FIGURE 4.3 The maximization of gross sustainable yield

Thus, the current annual increment equals the mean annual increment less mean annual planting costs. Since the current annual increment is a decreasing function of time,

$$T_N > T_G$$

We now turn to a comparison between the rotation period that follow from the Faustmann-Pressler-Ohlin (FPO) problem and T_N. The first order condition for the FPO rotation period in (4.4) can be rewritten

$$f'(T) = rf(T) + \frac{r[f(T)\,e^{-rT} - wp^{-1}l]}{1 - e^{-rT}}$$

$$= \frac{rf(T) - rf(T)\,e^{-rT} + rf(T)\,e^{-rT} - rwp^{-1}l}{1 - e^{-rT}}$$

$$= \frac{rf(T) - rwp^{-1}l}{1 - e^{-rT}} \tag{4.30}$$

However, when r tends to zero the right hand side of (4.30) tends to zero/zero, which is meaningless from a mathematical standpoint. L'Hospital's rule can be used to evaluate limits under such circumstances.[28] (For readers not familiar with l'Hospital's rule, the result

[28] $\lim_{r \to 0} h(r)/g(r) = \lim_{r \to 0} h'(r)/g'(r)$. A reader more familiar with Taylor's formula could use a Taylor expansion to derive the limits for $r \to 0$.

will be derived in section 5.5 by making use of a more direct method.) From l'Hospital's rule,

$$\lim_{r \to 0} \frac{rf(T) - rwp^{-1}l}{1 - e^{-rT}} = \lim_{r \to 0} \frac{f(T) - wp^{-1}l}{Te^{-rT}}$$

$$= f(T)\,T^{-1} - \frac{w}{p}\,lT^{-1} \tag{4.31}$$

which equals the right hand side of (4.29). Hence, when r is equal to zero, the FPO rotation period coincides with the rotation period which maximizes net sustainable yield, i.e.[29]

$$T_{\mathrm{N}} = T^0 \quad \text{for } r = 0 \tag{4.32}$$

Moreover, it is possible to prove that the FPO rotation period is shorter than the net sustainable yield maximizing rotation period for all T^0 and $r > 0$. To see this, return to the discussion concerning the influence of an increased interest rate on the FPO rotation period, where it was shown that $\partial T^0/\partial r < 0$ for all practically feasible cases.

We are now ready to sum up the findings of this section:

Proposition 4.4
The rotation periods T^0, T_{H}, T_{G}, T_{N} and T_{B} are related to each other in the following manner for $r \geqslant 0$:

(a) $T_{\mathrm{H}} > T^0$

(b) $T_{\mathrm{G}} < T_{\mathrm{N}}$

(c) $T_{\mathrm{B}} > T_{\mathrm{H}}$ if $\max r_{\mathrm{B}} < r$

$T_{\mathrm{B}} \leqslant T_{\mathrm{H}}$ if $\max r_{\mathrm{B}} \geqslant r$

(d) $T^0 < T_{\mathrm{N}}$ for $T^0, r > 0$

If $r \to 0$ one has

(e) $\lim_{r \to 0} T^0 = T_{\mathrm{N}}$

(f) $\lim_{r \to 0} T_{\mathrm{H}} = a \quad [f'(a) = 0]$

[29] The result is correct, but is not formally valid, since land value is not defined when $r = 0$. For an objection free derivation see section 5.5 below.

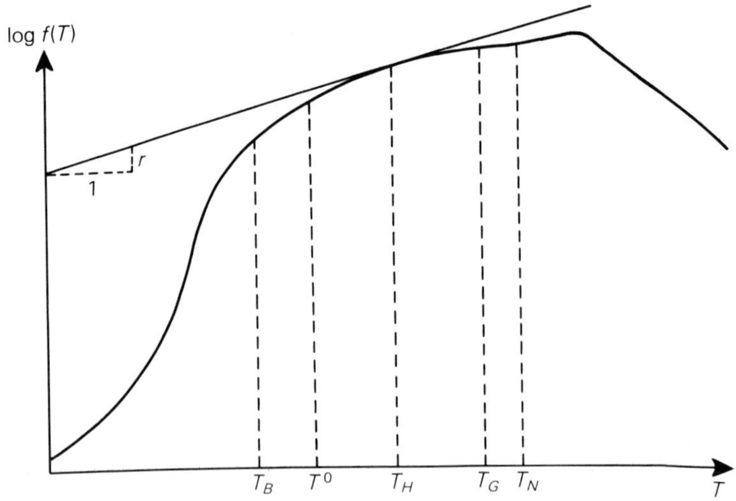

FIGURE 4.4 A possible rotation period configuration

A possible configuration of rotation periods is depicted in figure 4.4. Note that $\log f(T)$ is on the vertical axis of figure 4.4, and $\mathrm{d}\log f(T)/\mathrm{d}T = f'(T)/f(T)$. Hence we can determine the rotation period T_{H} by the graphical solution method already familiar from chapter 2.

5

A Further Analysis of the Optimal Rotation Period

5.1 Introduction

This chapter contains some important extensions of the problems discussed in chapter 4. We start by considering the effects of different kinds of taxes on the FPO rotation period and on the value of the forest land. Next, we consider the effect of both once and for all and continuous improvements in the biotechnology on the FPO rotation period (sections 5.3 and 5.4 respectively). In section 5.5 we extend the analysis of the determination of the optimal rotation period in two directions. Firstly, we consider the effects on the optimal rotation period when the price of wood and labour (regeneration costs) grow at a rate $a > 0$. Secondly, we analyse how the optimal rotation period can be determined when the real interest rate $r' = (r - a)$ is less than zero, i.e. when the infinite series of net revenues no longer converges. The chapter ends with a few comments on modifications of the restrictive assumptions upon which the discussions in chapters 4 and 5 rest.

5.2 The Effects of Taxes on the Rotation Period

The results in propositions 4.1 and 4.2 can be used to answer questions about how the optimal rotation period and the present value of land are affected by different kinds of taxes. A *sales tax* t on all future revenues from forestry is a decrease in the net price the forest owner obtains. Modifying (4.3), one obtains

$$V = [p(1-t)f(T)\,\mathrm{e}^{-rT} - wl](1 + \mathrm{e}^{-rT}\ldots) \tag{5.1}$$

The optimal rotation period increases since $-\partial T^0/\partial p > 0$ from proposition 4.1, while the land value V decreases since $-\mathrm{d}V/\mathrm{d}p = -\partial V/\partial p < 0$ from proposition 4.2.

A *payroll tax* α has a similar effect since it can be interpreted as an increase in the wage rate:

$$V = [pf(T)\,\mathrm{e}^{-rT} - w(1+\alpha)\,l\,](1 + \mathrm{e}^{-rT}\dots) \tag{5.2}$$

We know from proposition 4.1 that $\partial T^0/\partial w > 0$. From proposition 4.2 it follows that $\mathrm{d}V/\mathrm{d}w = \partial V/\partial w < 0$.

A *proportional profit tax* at the rate $0 < \theta < 1$ will not affect the optimal rotation period since

$$\begin{aligned}(1-\theta)\,V &= [p(1-\theta)\,f(T)\,\mathrm{e}^{-rT} - w(1-\theta)\,l\,](1 + \mathrm{e}^{-rT}\dots)\\ &= (1-\theta)[pf(T)\,\mathrm{e}^{-rT} - wl\,](1 + \mathrm{e}^{-rT}\dots)\end{aligned} \tag{5.3}$$

and $T^0(\lambda p, \lambda w, r) = T^0(p, w, r)$ from proposition 4.1. This claim follows by noting that λ can be set equal to $(1-\theta)$. This property can be more generally expressed by saying that the optimal rotation period is homogeneous of degree zero in prices, i.e. only relative prices matter. In turn, this property means that a general profit tax at the rate θ will decrease the value of land V to $(1-\theta)\,V$.

We now model the effect on the rotation period of two different kinds of *capital gains taxes*, one of which is calculated on a realization basis and the other on an accrual basis. The first case is simple. The capital gains tax on a *realization basis* is modelled by assuming that the forest owner pays a tax τ when the value of the trees increases from wl to $pf(T)$. Thus as the owner plants another tree, the net cash flow at time T is[1]

$$[pf(T) - wl\,](1-\tau) \tag{5.4}$$

and the present value of the future stream of net revenues can be written

$$\begin{aligned}V(T;\tau) &= -wl + \mathrm{e}^{-rT}\{[pf(T) - wl\,](1-\tau)\}\\ &\quad + \mathrm{e}^{-r2T}\{[pf(T) - wl\,](1-\tau)\}\dots \\ &= -wl + \frac{pf(T) - wl}{\mathrm{e}^{rT} - 1}(1-\tau)\\ &= \frac{[pf(T)\,\mathrm{e}^{-rT} - wl\,](1-\tau)}{1 - \mathrm{e}^{-rT}} - wl\tau\end{aligned} \tag{5.5}$$

A comparison of (4.3) and (5.5) shows that the rotation period which maximizes (4.3) also maximizes (5.5). In other words,

[1] $pf(T) - [pf(T) - wl\,]\,\tau - wl = [pf(T) - wl\,](1-\tau) + wl - wl$

provided that planting costs wl are constant over time, the introduction of capital gains taxation on a realization basis does not affect the optimal rotation period. Note, however, if one compares the effects on land values of a proportional profit tax and a capital gains tax with equal rates, i.e. $\theta = \tau$, the land value is lower under the capital gains tax. This result follows from a direct comparison of (5.5) and (5.3).

We now move to an analysis of the capital gains tax on an *accrual basis*.[2] The increase in value from t to $t + \Delta t$ is taxed at rate τ. Let $R(t)$ be the after tax value of a forest stand of age t. In a perfect market, where increases in net value are taxed at the accrual tax rate τ, the after tax value of the stand must, for all t, obey

$$\begin{aligned} R(t) &= \{R(t + \Delta t) - [R(t + \Delta t) - R(t)]\,\tau\}\, \mathrm{e}^{-r\Delta t} \\ &= [R(t + \Delta t)(1 - \tau) + \tau R(t)]\, \mathrm{e}^{-r\Delta t} \end{aligned} \tag{5.6}$$

Equation (5.6) simply says that the after tax value of the stand at t equals the discounted value of the after tax net return at $t + \Delta t$. The right hand side of (5.6) is a function of Δt. By a Taylor expansion[3] at $\Delta t = 0$ and ignoring higher order terms, one obtains

$$R(t) \approx R(t) - rR(t)(1 - \tau)\,\Delta t + R'(t)(1 - \tau)\,\Delta t - r\tau R(t)\,\Delta t$$

or

$$rR(t)\,\Delta t = R'(t)(1 - \tau)\,\Delta t$$

which can be rewritten as

$$\frac{R'(t)}{R(t)} = \frac{r}{1 - \tau} \tag{5.7}$$

This is a first order differential equation in t which has the solution

$$R(t) = A \exp\left(\frac{r}{1 - \tau}t\right) \qquad A = R(0)$$

where A is a constant to be determined from an initial or terminal condition. A terminal or boundary condition, if the stand is cut at $t = T^{\mathrm{a}}$, is

$$R(T^{\mathrm{a}}) = A \exp\left(\frac{r}{1 - \tau}T^{\mathrm{a}}\right) = pf(T^{\mathrm{a}})$$

[2] The analysis follows Kovenock and Rothschild (1983).

[3] For Taylor's formula, the reader is referred to any introductory book on calculus.

Hence, the market value of the stand cut at $t = T^a$ at $t = 0$ is

$$A = pf(T^a)\exp\left(-\frac{r}{1-\tau}T^a\right)$$

If planting costs are wl, a plot of land on which trees are planted at $0, T^a, 2T^a \ldots$ yields returns equal to

$$PV = \frac{pf(T^a)\exp\left(-\dfrac{r}{1-\tau}T^a\right) - wl}{1-\exp(-rT^a)} \tag{5.8}$$

One can establish that the optimal cutting time for trees which are taxed on an accrual basis is a non-increasing function of the tax rate. To prove this we set

$$r' = \frac{r}{1-\tau}$$

and

$$dr' = \frac{r\,d\tau}{(1-\tau)^2} > 0$$

i.e. an increase in the tax rate τ increases r'. This means that to prove the assertion, it is sufficient to prove that $\partial T^a/dr' < 0$. The first order conditions for an interior maximum can be written

$$pf'(T^a) - r'pf(T^a) - rPV = 0$$

Differentiation with respect to r' yields (omitting, for convenience, the superscript a on the rotation period T^a)

$$[pf''(T) - r'pf'(T)]\,dT = \left[pf(T) + r\frac{\partial PV}{\partial r'}\right]dr' \tag{5.9}$$

where

$$r\frac{\partial PV}{\partial r'} = -rT\frac{pf(T)\,e^{-r'T}}{1-e^{-rT}} < 0$$

The right hand side of (5.9) can be written

$$\frac{pf(T)(1-e^{-rT}-rT\,e^{-r'T})}{1-e^{-rT}}$$

This expression is non-negative provided that

$$H(T) = (1-e^{-rT}-rT\,e^{-r'T})$$

is non-negative. Clearly $H(0) = 0$, and

$$\frac{\partial H}{\partial T} = H'(T) = r\,e^{-rT} + rr'T\,e^{-r'T} - r\,e^{-r'T} \geqslant 0 \quad \text{all } T > 0$$

since $r' \geqslant r > 0$. This means that $H(T) \geqslant 0$ for all T. Hence, the right hand side of (5.9) is non-negative. Combined with the fact that $pf''(T) - r'pf(T) < 0$, the assertion that $\partial T^a/\partial r' \leqslant 0$ follows. For $T^a > 0$, i.e. for an interior solution, $\partial T^a/dr' < 0$ for $r > 0$.

We end this section on taxes by considering an annual lump sum tax, which is related to the value of the land either indirectly through the site quality classification index of the forest land – the annual sum increases with the quality of the forest land – or by a direct relation between the annual lump sum and the value of the forest land.

In the former case, the land value after tax can be written

$$\max_T V(T; a) = \max_T \left[V(T) - \frac{a(i)}{r}\right] \tag{5.10}$$

where $a(i)$ is the annual lump sum as a function of the site index of the land, and $a(i)/r$ is the present value of the perpetuity $a(i)$. It should be obvious from (5.10) that after tax value is maximized when $V(T)$ is maximized. Hence, the optimal rotation period under a lump sum, $a(i)$, does not deviate from the FPO rotation period in the no tax case.

An annual lump sum b which is directly related to the value of the forest land can be modelled by assuming that it is based on the rule that a certain proportion of land value β should be delivered in taxes. The present value of the lump sum payments equals β times the present value of the forest land:

$$\int_0^\infty b\,e^{-rt}\,dt = \beta V(T) \qquad 0 < \beta < 1$$

Solving for b, one obtains

$$b = r\beta V(T)$$

and the after tax land value can be written

$$\max_T V(T, \beta) = \max_T\,(1 - \beta)\,V(T) \tag{5.10a}$$

which means that the effect of the lump sum tax $b(\beta, T)$ coincides with the effect of a proportional tax at the rate $\theta = \beta$.

We now summarize our discussion of the effects of different kinds of taxes on the rotation period and the value of land. The main points are:

Proposition 5.1
If forestry income is taxed, the following statements are valid:

(a) A sales tax will lengthen the optimal rotation period.[4]
(b) A payroll tax will lengthen the optimal rotation period.
(c) A proportional profit tax will not affect the optimal rotation period.
(d) A capital gains tax on a realization basis will not affect the optimal rotation period.
(e) A capital gains tax on an accrual basis will shorten the optimal rotation period.
(f) A lump sum tax, i.e. a sum of money that the forest owner has to pay every year independently of whether he cuts or not, will not affect the optimal rotation period.[5]
(g) All kinds of taxes will decrease the value of land.

5.3 The Effects of a Once and For All Change in the Biotechnology

As mentioned in chapter 3, it is possible to model changes in biotechnology by introducing a biological parameter into the growth function. In addition, it is necessary to specify how changes in the parameters change the properties of the growth function. We start by analysing a particular parameterisation of the growth function:[6]

$$f(T, \alpha) = \alpha f(T) \qquad \alpha > 0 \tag{5.11}$$

An increase of α means that the amount of the biomass is changed proportionally for all T, i.e. a doubling α doubles the biomass. This is a *time neutral change of technology* (TNCT), where technology here refers to biotechnology. 'Neutral' is used to indicate that the rate of current annual increment is independent of biotechnological progress. A TNCT method is, in most instances, too simplified. Consider, for example, a switch from *Pinus silvestris* to *Pinus*

[4] The optimal rotation period here refers to the optimal rotation period according to Faustmann in the case of no taxes.
[5] Unless its size is large enough to make the harvest unprofitable (the value of the land negative). In that case, the stand will never be cut.
[6] This section is partly based on Löfgren (1984b).

contorta. From a study of the growth functions of these two species, one knows that the growth of *Pinus contorta* is faster but that the relative difference declines over time.[7] This may be modelled by letting $\alpha > 1$ be a decreasing function of time for some $T > \bar{T}$. There are also differences in value between a *Pinus contorta* tree and a *Pinus silvestris* tree. The value of a cubic metre of *Pinus contorta* is probably lower than the value of a cubic metre of *Pinus silvestris*, owing to the fact that *Pinus contorta* is less suitable for lumber production. Since value growth is the relevant factor for the determination of the optimal rotation period, information on differences in volume growth functions is in general insufficient to determine differences in the optimal rotation periods.[8] One must in addition have information on the prices of different species of wood. Some of these complications will be dealt with in the next section.

The treatment of the effects of a TNCT on the optimal rotation period is simple. An increase of α is equivalent to a once and for all increase in the price (stumpage value). From proposition 4.1 we know that a price increase reduces the optimal rotation period. Hence, the same holds true for a TNCT improvement.

One may consider the effects of a more efficient cutting technology on the optimal rotation period by using the results in proposition 4.1. A more efficient cutting technology increases stumpage values or the net price per cubic metre and results in a shortening of the optimal rotation period. Similarly a more efficient planting technology is equivalent to a lower wage rate and, hence, implies a shortening of the optimal rotation period. Another conclusion is that the value of the forest land increases with respect to all considered changes in technology, holding all other factors constant.

The situation becomes more complicated when the biotechnologies of the standing timber and newly seeded land are different. Equation (4.12) remains valid, although the optimal rotation period of the standing timber T^s in general will differ from the optimal rotation period of the infinite series created by the rotations on the newly seeded land $T(\alpha)$. If the new technology means a once and for all genetically induced TNCT improvement, then $T^s < T^0$, where T^0 is the rotation period on newly seeded land under the old technology. This is because higher land values caused by the new technology result in increases in the interest on the value of bare land. This in turn means that relative value growth in the new

[7] Max $[f(T)/T]$, i.e. culmination is earlier for *Pinus contorta*, which shows that the differences in the growth functions cannot be a TNCT.

[8] By assuming $p =$ constant, we have in addition ducked the problem that the value of a tree depends on its age. On the other hand, very little of generality is lost by this simplification.

optimum has to be higher, and this can only be accomplished through a shortening of the rotation period.[9]

Finally, consider the effect of a TNCT improvement on the value of forest land. The value of bare land under the α biotechnology can be written

$$V_\alpha = V_\alpha[T_\alpha(\alpha p, w, r), \alpha p, w; r] \tag{5.12}$$

The change in the value of forest land from an increase in α is

$$\frac{\mathrm{d}V_\alpha}{\mathrm{d}\alpha} = \frac{\partial V_\alpha}{\partial T_\alpha}\frac{\partial T_\alpha}{\partial \alpha} + \frac{\partial V_\alpha}{\partial \alpha} = \frac{\partial V_\alpha}{\partial \alpha} = \frac{pf(T_\alpha)}{(\mathrm{e}^{rT}\alpha) - 1} \tag{5.13}$$

since $\partial V/\partial T_\alpha = 0$ when evaluated at optimum. The interpretation of (5.13) is that the total derivative of land value with respect to α equals the discounted value of the yield of all future harvests obtainable from the original technology, when the future harvests take place at T_α, $2T_\alpha, \ldots$. Formula (5.13) is handy when it comes to assessing the potential economic gain from genetical progress.

5.4 The Effects of Regular Improvements in Biotechnology

A related problem is the effects on the optimal rotation periods caused by regular TNCT improvements. The α in (5.11) is now allowed to change over time. The biotechnology improvement may be caused, for example, by regularly improving genetic material.[10] This can be modelled by letting $f(T)$ be the growth function during the first rotation. Genetic improvements are assumed to follow a multiplicative TNCT pattern of the form

$$1,\ \phi_1(T_1),\ \phi_1(T_1)\,\phi_2(T_2),\ \phi_1(T_1)\,\phi_2(T_2)\,\phi_3(T_3), \ldots$$

generating the growth functions

$$f(T_1),\ \phi_1(T_1)\,f(T_2),\ \phi_1(T_1)\,\phi_2(T_2)\,f(T_3),$$
$$\phi_1(T_1)\,\phi_2(T_2)\,\phi_3(T_3)\,f(T_4)\ \ldots$$

[9] In the problem specified in (4.12), a once and for all improvement of the biotechnology would result in a first order condition which can be written

$$\frac{f'(T)}{f(T)} = r + r\frac{V[T(\alpha)]}{pf(T)} \qquad V[T(\alpha)] > V(T^0)$$

and if T is chosen equal to T^0, then the right hand side would be larger than the left hand side. To obtain equality T must be chosen equal to some $T^S < T^0$.

[10] By genetically improved material we simply mean a TNCT improvement of the growth function. No qualities like robustness or resistance to attacks from insects are implied.

Further assume that:

$$\phi_1\phi_2,\ldots,\phi_i \leqslant \phi_1\phi_2,\ldots,\phi_i\phi_{i+1} \qquad \text{all } T, \text{ all } i \tag{5.14}$$

or

$$\phi_i \geqslant 1 \qquad \text{all } i$$

Equation (5.14) means that genetic progress is a multiplicative process and monotone. Finally, assume that there is a physical upper bound on the possible genetic improvement of a species,[11] i.e

$$\lim_{n\to\infty} \prod_{i=1}^{n} \phi_i(T_i) = \bar{B} < \infty \tag{5.15}$$

The present value of the stream of future net revenues along a cutting programme $T_1, T_2, T_3, \ldots, T_\infty$ can be written (assuming $p = 1$, and planting costs equal zero)

$$\begin{aligned} PV = V_1 = {} & f(T_1)\exp(-rT_1) + \phi_1(T_1)f(T_2)\exp[-r(T_1+T_2)] \\ & + \phi_1(T_1)\phi_2(T_2)f(T_3)\exp[-r(T_1+T_2+T_3)] + \ldots \\ & + \left[\prod_{i=1}^{n} \phi_i(T_i)\right] f(T_{n+1})\exp -r\sum_{i=1}^{n+1} T_i + \ldots \end{aligned} \tag{5.16}$$

which can be rewritten as

$$V_1 = f(T_1)\,e^{-rT_1} + \phi_1(T_1)\,e^{-rT_1}\,V_2(T_2,\ldots;r) \tag{5.16a}$$

where

$$V_2 = f(T_2)\,e^{-rT_2} + \phi_2(T_2)f(T_3)\,e^{-(T_2+T_3)}\ldots$$

is independent of T_1. V_2 is the value of bare land at $T = T_1$, when the biotechnology is improved according to the multiplicative TNCT pattern

$$1,\ \phi_2(T_2),\ \phi_2(T_2)\,\phi_3(T_3),\ldots \tag{5.17}$$

and rotations $T_2, T_3, \ldots$ are chosen. The reader should note that (5.14) and (5.17) are different since the biotechnological improve-

[11] Note that this implies that

$$\lim_{n\to\infty} \phi_n(T_n) = 1 \text{ as } \prod_{i=1}^{n}\phi_i = \phi_n \prod_{i=1}^{n-1}\phi_i, \text{ where } \prod_{i=1}^{n}\phi_i \text{ equals } \phi_1\phi_2\ldots\phi_n.$$

ment during the first rotation period, $\phi_1(T_1)$, is eliminated from the TNCT pattern (5.17). In particular, define

$$V_2^* = V_2^*[T_2^*(r), T_3^*(r), \ldots; r] = \max_{T_i;\, i \in I} V_2(\cdot) \quad I = \{T_2, T_3 \ldots\}$$

as the land value under the particular TNCT pattern in (5.17).

More generally,

$$V_n^* = V_n^*(r)$$

is the land value under the TNCT pattern

$$1,\ \phi_n(T_n^*),\ \phi_n(T_n^*)\,\phi_{n+1}(T_{n+1}^*), \ldots \tag{5.18}$$

Invoking Bellman's principle of optimality, we can 'solve' for T_1 by assuming that $V_2^*(r)$ is known. Maximize

$$V_1 = e^{-rT_1}[f(T_1) + \phi(T_1)\, V_2^*(r)]$$

with respect to T_1. The first order condition for a maximum can be written

$$\frac{f'(T_1^*)}{f(T_1^*)} = r + [r\phi_1(T_1^*) - \phi_1'(T_1^*)]\,\frac{V_2^*}{f(T_1^*)} \tag{5.18a}$$

where $r - \phi_1'/\phi_1$ is the real interest rate under the first rotation. From Bellman's principle, we know that the first order condition for rotation i is

$$\frac{f'(T_i^*)}{f(T_i^*)} = r + [r\phi_i(T_i^*) - \phi_i'(T_i^*)]\,\frac{V_{i+1}^*}{f(T_i^*)} \tag{5.18b}$$

In combination with an assumption that $r\phi(T_i) - \phi_i'(T_i) \geqslant r$ all i, (5.18) can be used to show that the T_i are shorter than the FPO rotation period, T^0, which is obtained under the TNCT process

$$1,\ 1,\ 1, \ldots$$

i.e. when there is no improvement of the biotechnology. This is the case because, among other things, $[r\phi_i(T_i^*) - \phi_i'(T_i^*)]\, V_{i+1} > rV(T^0)$. At the limit, however, genetic progress ceases, i.e.

$$\lim_{n \to \infty} \phi_n(T_n) = 1$$

Moreover (5.15), in conjunction with Cauchy's principle of convergence, implies that

$$\lim_{n \to \infty} \prod_{i=n}^{\infty} \phi_i = 1$$

Hence

$$\lim_{n \to \infty} [r\phi_n(T_n^*) - \phi_n'(T_n^*)] V_n = rV(T^0)$$

and

$$\lim_{n \to \infty} T_n = T^0$$

To compare T_1^* with T_i^* we need information on the magnitudes of $\phi_1(T_1^*) V_1^*$, $\phi_i(T_i^*) V_{i+1}^*$ and $\phi_1'(T_1^*)$, $\phi_1'(T_i^*)$, where ϕ_i' is the marginal rate of improvement of the biotechnology over time during the ith rotation. It is a reasonable, but nevertheless *ad hoc*, hypothesis that

$$\phi_i V_{i+1}^* - \phi_{i-1} V_i^* > 0 \qquad \text{for all } i \leqslant \text{some } n'$$

and (5.19)

$$\phi_i V_{i+1}^* - \phi_{i-1} V_i^* \leqslant 0 \qquad \text{for } i > n'$$

This assumption means that for $i \leqslant n'$ one faces better and better TNCT patterns, while as biotechnological progress approaches the upper bound $\bar{B}$ it slows down and the TNCT patterns deteriorate over time.

Given the assumption explicit in (5.19), and neglecting the magnitude of the ϕ_i', the rotation period pattern under multiplicative, regular TNCT improvements resembles that in figure 5.1. The

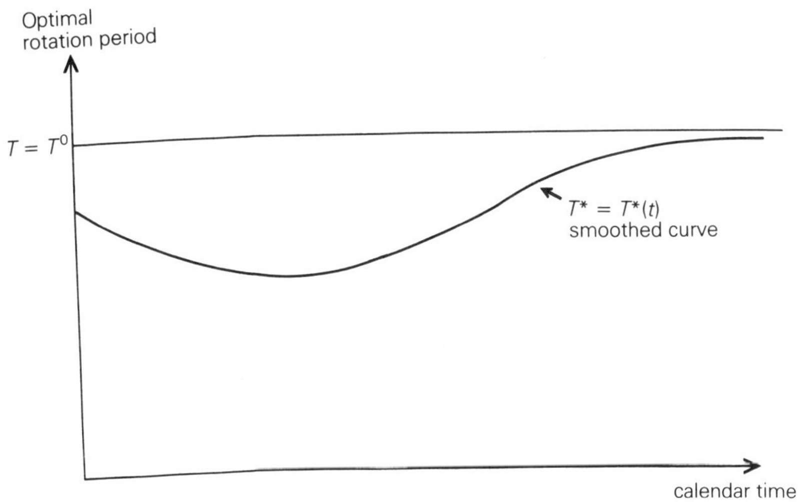

FIGURE 5.1 The optimal rotation period pattern under a multiplicative TNCT improvement

optimal rotation period under the TNCT approaches the FPO period under a constant biotechnology, since the rate of increase in value,

$$\prod_{i=1}^{n-1} \phi_i f'(T_n) \Big/ \prod_{i=1}^{n-1} \phi_i f(T_n) = f'(T_n)/f(T_n)$$

is independent of techological progress under a TNCT. Moreover, the real interest on the land value, deflated by the value of the stand, is asymptotically independent of biotechnological progress. This property is due to the seemingly innocent assumption that regeneration costs are equal to zero. With positive regeneration costs, the long run TNCT rotation period would be shorter than the FPO period. Genetic progress at a constant price is equivalent to an increase in stumpage values. When regeneration costs are positive, a price increase lowers regeneration costs relative to the price of the wood (real regeneration costs w/p) and this shortens the optimal rotation period as seen in proposition 4.1.

The main conclusion which follows from the analysis is that future TNCT progress, not its current level, determines whether the FPO with such progress is shortened by comparison with the FPO without it. Put differently, it is not the improved growth, but the known possibilities of future improvements of the growth function, which cause a shortening of the optimal rotation period.

All comparisons assume that all other factors (such as the net price and the interest rate) are constant. In addition, we assumed that biotechnological progress was neutral, which means that the current rate of annual increment $f'(T)/f(T)$ was independent of the level of the realized progress. Whether biotechnological progress is neutral or not is an important empirical question, which we have so far avoided in this book. However, from our results – owing to the choice of TNCT as a norm – we can conclude that if biotechnological progress improves the rate of current annual increment, this progress would *ceteris paribus* tend to lengthen the FPO.[12] The opposite holds if biotechnological progress cuts down on the rate of current annual increment.

The analysis has been conducted at the stand level. The analysis should apply to the farm level provided that the only difference between the latter and the stand level is the extent to which scale effects are present (non-linearities and/or capacity constraints),

[12] The left hand side of (5.18) increases, while the right hand side remains unchanged owing to the *ceteris paribus* clause. From the strict concavity of the present value function it then follows that the rotation period must be increased to restore the equality. The *ceteris paribus* clause is, however, very restrictive.

given that the non-linearities are negligible. If there are increasing returns to scale in the cutting technology owing to a higher stocking level generated by biotechnological progress, this would tend to shorten the optimal rotation period (*ceteris paribus*); decreasing cutting costs per cubic metre are equivalent to an increased net price, which according to proposition 4.1 shortens the optimal rotation period.

The results obtained in this section imply that a shortening of the rotation period must be motivated not solely by a previously experienced improvement in biotechnology, but also by the certain knowledge of future improvements.[13] This hints at the difficult problems caused by an imperfect knowledge of the future. Given that biotechnological progress and the robustness to external shocks of the improved growth function are uncertain, how does one determine the FPO in the presence of biotechnological progress and how does this compare with the FPO in the absence of progress? This topic, however, is outside the scope of this book.

5.5 A Slightly More General Setting

In this section we slightly extend the analysis of chapter 4 in two directions. Firstly, we show how the optimal rotation period is affected by the fact that the prices of wood and labour grow at a rate[14] $a \neq 0$, i.e.

$$\begin{aligned} \dot{p} &= ap \qquad p(0) = p_0 \\ \dot{w} &= aw \qquad w(0) = w_0 \end{aligned} \tag{5.20}$$

Secondly, we show how the rotation problem can be solved when the real interest rate is negative.

If inflation obeys the differential equations (5.20), one obtains the price and wage paths as

$$\begin{aligned} p(T) &= p_0 \, \mathrm{e}^{aT} \\ w(T) &= w_0 \, \mathrm{e}^{aT} \end{aligned} \tag{5.21}$$

and the value of forest land can be written

$$\begin{aligned} V = &\{p_0 f(T_1) \exp[(a-r)T_1] - w_0 l\} \\ &+ \{p_0 f(T_2) \exp[(a-r)T_1] - w_0 l\} \exp[(a-r)T_2] \end{aligned}$$

[13] We are here assuming negligible regeneration costs.

[14] We will only consider the case $a > 0$, but the reader could easily work out the details for the case $a < 0$.

$$+ \ldots + \{p_0 f(T_n) \exp[(a-r) T_1] - w_0 l\}$$

$$\times \exp\left[(a-r) \sum_{i=2}^{n} T_i\right] + \ldots \tag{5.22}$$

We did not formally show, when the standard rotation problem was solved, that the optimal rotation period must be a constant. This is, however, easily understood since the optimization problem facing a forester at time T_n is the same problem that he faces at times T_{n-1}, up to a positive multiplicative constant.

The generalized rotation problem can be written

$$\max_T V = \frac{p_0 f(T) \mathrm{e}^{-r'T} - w_0 l}{1 - \mathrm{e}^{-r'T}} \qquad r' = r - a > 0 \tag{5.23}$$

where $r - a$ is the real rate of interest, provided that all prices in the economy increase at rate a. If a only measures the rate at which forestry prices change, this interpretation is no longer possible. Then $r - a$ might be interpreted as the 'real interest rate' of the landlord since he has the choice of either cutting down the forest and earning interest on the proceeds or maintaining the forest stand and earning growth plus the relative price increase of the timber. Clearly, this does not change very much so long as $r' > 0$. Formula (5.23) is nothing but a more generalized version of the FPO problem.

As shown in proposition 4.1, the effect of a decrease in the interest rate on the rotation period is positive. This also holds for the effect of the inflation rate on the rotation period.[15] Although the real wage rate facing the firm is kept constant over time ($w/p = w_0/p_0$) by the assumed inflation pattern, the rate of inflation still affects the optimal rotation period through its effect on the real interest rate.

Zero or negative interest rates may seem very unrealistic. Regarding realism, however, it is the real interest rate that matters, and a rate of inflation larger than the nominal interest rate ($a > r$) means that $r'' = r - a < 0$. A zero or negative interest rate means that the infinite series created by the FPO problem does not converge, and the value of land is infinite. In a competitive economy, however, one can always obtain the maximum yearly amount that the market is willing to pay for the option to rent the land for forestry. Formally, this amount of money will depend on the rotation period T and emerges as the solution to the problem

$$\max_T R(T)$$

[15] The factor a is what the early German authors called *Teuerungszuwachs*.

subject to

$$pf(T)\,\mathrm{e}^{-rT} - wl - R\int_0^T \mathrm{e}^{-rt}\,\mathrm{d}t = 0 \tag{5.24}$$

where R is the land rent in the interval $\mathrm{d}t$.

This formulation implicitly assumes that the market squeezes the maximum willingness to pay from the people who rent the land. Solving (5.24) for R yields

$$\max_T R = \frac{r[pf(T)\,\mathrm{e}^{-rT} - wl]}{1 - \mathrm{e}^{-rT}} \tag{5.25}$$

For $r > 0$, R is the interest income on the value of land or, in terms of capital theory, is just income. For a given $r > 0$, the maximization of the value of land is equivalent to the maximization of interest income on the value of land.[16] Hence, the T that solves (5.25) coincides with the FPO rotation period.

This interpretation does not hold when $r < 0$, but the method to determine the optimal rotation periods works. The first order condition for the case $wl = 0$ can be written

$$\frac{f'(T^0)}{f(T^0)} = r + \frac{r}{\mathrm{e}^{rT^0} - 1} = \frac{r\,\mathrm{e}^{rT^0}}{\mathrm{e}^{rT^0} - 1} = \frac{r}{1 - \mathrm{e}^{-rT^0}} > 0 \tag{5.26}$$

The first order condition implies that the resulting optimal rotation period does not mean that the forest stand is decaying. Since the right hand side of (5.26) is positive for $r \neq 0$, it must hold at an optimum that $f'(T^0) > 0$.

The limit of the right hand side of (5.26) exists for $r = 0$ but it is also possible to proceed in the following manner:

$$\max_T R(T)$$

subject to

$$pf(T) - R\int_0^T 1\,\mathrm{d}t = 0 \tag{5.27}$$

which results in the maximization of the gross sustainable yield. Thus, if one disregards planting costs, the FPO rotation period for

[16] If x^* maximizes $f(x)$ it also maximizes the function $rf(x)$, where r is a constant larger than zero.

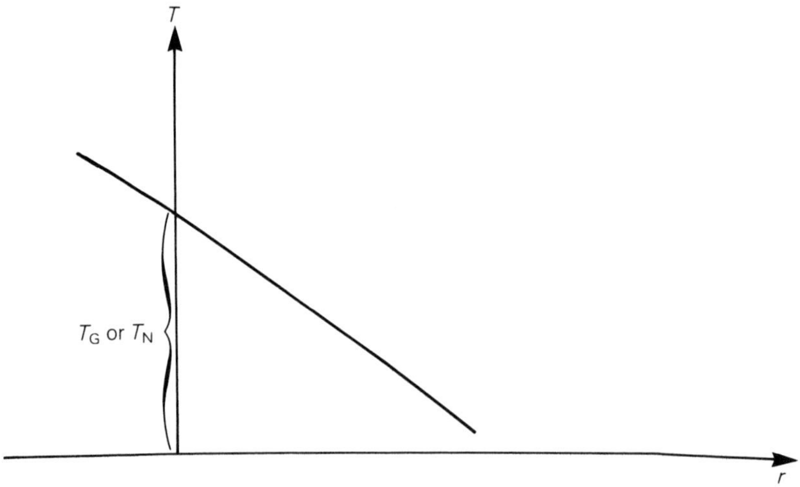

FIGURE 5.2 The FPO rotation period as a function of the interest rate. $T_N(T_G)$ is the maximum net (gross) sustainable yield rotation period

$r = 0$ coincides with the maximum gross sustainable yield rotation period[17] T_G $[f'(T_G) = f(T_G)/T_G]$. By showing that

$$\frac{1}{T_G} - \frac{r}{1 - e^{-rT_G}}$$

is larger than zero for $r < 0$, it follows from decreasing returns to scale $(f'' < 0)$ that the FPO rotation period is longer than the maximum gross sustainable yield rotation period T_G for $r < 0$. The reader is invited to work out the details. This result means that one can depict the FPO rotation period as a continuous and decreasing function of the real interest rate (see figure 5.2).

The reader should also note that, if $r < 0$, according to the wine aging rule it is optimal to save the forest stand until it decays, i.e. until $f'(T_H) < 0$. This is, as has just been noted, never true for the FPO rotation period.

5.6 Concluding Comments

Chapters 4 and 5 on the rotation problem rely on many specific assumptions, and it is desirable to relax some of them. Perhaps one

[17] Given an account of planting costs, the FPO rotation coincides with the maximum net sustainable yield rotation T_N $(> T_G)$ derived in (4.29).

of the most unrealistic assumptions is that the same relative prices prevail in the present as in the entire future. This leads to somewhat unexpected results on the long run supply of roundwood. The supply of roundwood from a stand when cut can be written

$$Q_{t+T^0} = f[T^0(p, w, r)] \tag{5.28}$$

Assume that the price of roundwood increases. According to proposition 4.1, the rotation period decreases for all existing stands and the long run supply of roundwood decreases since $f(T)$ is an increasing function of T. In the short run, however, one could expect an increasing supply since a once and for all increase of p makes many stands 'over aged' in the sense that the actual age is greater than the new FPO rotation period.

The backward bending long run supply curve is a substantial oversimplification in its assumptions that the amount of land used for forestry is independent of the profitability of forestry, and that the biotechnology is independent of the stumpage value. A better price for timber would result in more land being used for forestry and in the introduction of more advanced technologies. This should be sufficient to cause a normal, positively sloped, long run supply function.

A good way to phrase the message of this chapter is to say that (5.28) is not a supply function in the usual sense. A supply function tells us what happens to the quantity supplied when *ceteris paribus* the price increases in a particular period. To obtain such a supply function, one must relax the assumption of one price independent of time. This will be done in the next chapter. The FPO approach at the stand level also assumes that there are no restrictions on the age composition of the forest farm. If there are restrictions on the age composition of the forest farm, the FPO rotation period at the stand level will no longer be optimal.[18] This undesirable property will also be removed by the models of chapter 6. Finally, we ask the following important question. Under what conditions can the rotation problem be treated at the stand level, as in the present chapter, and when is it necessary to take account of the composition of the entire forest? In chapter 6 we show that a cutting technology with increasing or decreasing returns to scale means that the rotation problem can no longer be treated at the stand level.

[18] This is shown very convincingly in a recent study by Hultkrantz (1982), where it is shown that the optimal rotation period of the so-called synchronized steady state forest, with stands of equal area in each age from seeded land to maturity, has a maturity age shorter than the FPO rotation period. See also Nautiyal and Fowler (1980).

6

The Linear and Non-linear Forest and the Supply of Wood

6.1 Introduction

In this chapter we consider a more general forest management problem than the so-called Faustmann-Pressler-Ohlin problem considered in chapters 4 and 5. In section 6.2 we introduce a very general infinite planning horizon model, taking account of the age structure of the timber firm and the possibility that roundwood prices may vary between periods. This model is simplified in section 6.3, resulting in a conventional linear programming problem, i.e. maximizing a linear profit function subject to linear restrictions. Formal properties of the solution are derived in section 6.4. We also generalize the model to consider non-linear objective functions. This analysis is continued in section 6.5 where more specific properties of the timber supply function are derived. In section 6.6 we discuss the effects of different kinds of taxes on the supply of roundwood. The chapter ends with a discussion of some possible generalizations of the model. We introduce a cost function and discuss how the initial endowments affect timber supply.

6.2 A General Age Class Model

Consider a timber firm with a given amount of land $\bar{M}$, where the trees are of different ages.[1] Let

$$x = x(t, a) \tag{6.1}$$

[1] The particular continuous version of the model has been borrowed from Berck (1976). The discrete time linear model is also identical with the one in Berck (1976), but the

be the number of hectares at time t of timber in age class a. Furthermore, let Ω be the set of possible ages. By definition,

$$\int_{\omega \in \Omega} x(t, \omega)\, \mathrm{d}\omega = \bar{M} \tag{6.2}$$

The sum (the integral) of timber of different age classes covers the entire forest land. Timber felling is defined by

$$c = c(t, a) \tag{6.3}$$

i.e. c is the number of hectares with trees of age a, which is cut at time t. It also holds that

$$x(t, a) = x(t-s, a-s) - \int_0^s c(t-\omega, a-\omega)\, \mathrm{d}\omega \quad \text{all } s \tag{6.4}$$

The area where timber of age class a grows at time t equals the area that, at time $t-s$, was occupied by age class $a-s$ minus what was cut in the interval $[0, t-(t-s)] = (0, s)$. Finally,

$$\int_{\omega \in \Omega} c(t, \omega)\, \mathrm{d}\omega = x(t, 0) \tag{6.5}$$

i.e. the cut at time t equals the area occupied by age class zero at time t.

Equations (6.1)–(6.5) define the forest's technology. To formulate an economic management problem, one must have the volume of timber of age class ω at time t, and its price. Let

$$g = g(\omega) \tag{6.6}$$

be the amount of wood (in cubic metres) on a hectare of land of age class ω. In addition, assume that the price of any age class depends only on time and not on the age class. Given a perfect capital market, the forest owner's investment criterion is to maximize the present value of the income stream from cutting (this was shown in chapter 1). Hence, the forest management problem is formulated as

$$\max_{c(t,a)} \int_0^\infty [p(t)\, \mathrm{e}^{-rt} \int_{\omega \in \Omega} c(t, \omega)\, g(\omega)\, \mathrm{d}\omega]\, \mathrm{d}t \tag{6.7}$$

---

approach as such is not particular new. The Swedish Domän (The Forest Service) group has planned their cutting policy with a similar model in the 1960s, and Jungenfelt (1973) uses such a model. Versions with exogenous thinning can be found in Hultkrantz (1982) and Kallio (1982). For a closely related approach see also Lyon and Sedjo (1983) and Mitra and Wan (1982).

subject to the technology defined by (6.1)–(6.5) and some initial conditions. Here $p(t)$ is the price of one cubic metre of wood and r is the prevailing discount rate in a perfect capital market.

We do not intend to solve this problem. Its formulation, however, illustrates the complexity of the continuous infinite horizon approach. A similar, but somewhat less complicated problem, is solved very neatly in Heaps and Neher (1979) and Heaps (1984).

6.3 The Linear Forest

In this section we generate a linear programming problem to maximize the present value of the income stream from cutting, subject to the constraints of a 'linear' technology. Two simplifying assumptions are needed. First, assume a finite planning horizon. It should be noted that this assumption is not logical. The age distribution of the forest in any given future period has implications for the future. If one stops the analysis after a finite number of periods, say T, it is in principle necessary to include some 'scrap value' of the forest stands. The only logical way of doing so is to determine the maximum present value attainable in future periods, starting with a fixed age distribution.

On the other hand, since prices are bounded, it can be shown that the loss in present value from stopping after a finite number of periods can be made arbitrarily small by making T large. If prices are bounded and the amount of land is finite, then the maximum loss per period is bounded. Call this upper bound B. The maximum loss in present value by stopping at time T is given by

$$\int_T^\infty B\,\mathrm{e}^{-r(T+t)}\,\mathrm{d}t = \mathrm{e}^{-rT} B \int_T^\infty \mathrm{e}^{-rt}\,\mathrm{d}t = \frac{B\,\mathrm{e}^{-2rT}}{r} \tag{6.8}$$

which can be made smaller than any ϵ (provided that $r > 0$) by choosing T large enough.

The second simplifying assumption required is that time is discrete and not continuous. We now introduce the technology of the discrete time, finite horizon version of the model in (6.1)–(6.7). Let x_{ti} be the number of acres of land occupied at time t by trees of age i. This corresponds to (6.1). Moreover,

$$\sum_{i=0}^{n} x_{ti} = \bar{M} \quad \text{all } t \tag{6.9}$$

where n is the highest possible age class. Equation (6.9) corresponds to (6.2). In addition, let the harvest be a joint product with timber and seeded land as outputs.

We now define the discrete analogy to (6.4). Given a highest age class n, and lowest age class (seeded bare land) 0, bare land at the end of period 1 is

$$x_{10} = c_{11} + c_{12} + c_{13} \ldots + c_{1n} \tag{6.10}$$

where c_{1i} is the number of acres with trees of age i which is cut in period 1.

The area of land occupied by trees of age $i < n$ at time $t \leqslant T$ is equal to

$$x_{ti} = -c_{ti} + x_{t-1\,i-1} \quad \text{all } t, \quad i = 1, \ldots, n-1 \tag{6.11}$$

i.e. what was left from the previous period less the period's cut. For $i = n$ we assume, to end the recursion, that the previous period's remaining trees of age class n stay in this age class for ever. In other words,

$$x_{tn} = -c_{tn} + x_{t-1\,n-1} + x_{t-1\,n} \tag{6.12}$$

Combined with a growth function, this means that if the growth of the biomass of an even-aged stand ends after n periods, any extra growth is balanced by mortality. For all practical purposes, the assumption behind (6.12) does not constitute a major violation of reality. Finally, as an initial condition,

$$\bar{M} = \bar{x}_{00} + \bar{x}_{01} + \ldots + \bar{x}_{0n} \tag{6.13}$$

i.e. seeded land and trees of different age classes cover the initial amount of land. (One could allow for inequality by defining the amount of unseeded land and by defining c_{t0} as the harvest of unseeded land, which involves no output, but the seed as input.)

The restrictions (6.10)–(6.13) can be stated in matrix notation as

$$Ix(0) = \bar{x}(0)$$

$$Ix(t) - Bc(t) - Ax(t-1) = 0 \qquad t = 1, \ldots, T$$

where

$$x(t) = \begin{bmatrix} x_{t0} \\ \cdot \\ \vdots \\ x_{tn} \end{bmatrix} \qquad c(t) = \begin{bmatrix} 0 \\ c_{t1} \\ \vdots \\ c_{tn} \end{bmatrix}$$

and I, A and B are $(n+1) \times (n+1)$ matrices:

$$I = \begin{bmatrix} 1 & 0 & \cdots & 0 \\ 0 & 1 & & \vdots \\ \vdots & & \ddots & \\ \vdots & & 1 & 0 \\ 0 & \cdots & 0 & 1 \end{bmatrix} \quad A = \begin{bmatrix} 0 & 0 & \cdots & & 0 \\ 1 & 0 & & & \vdots \\ 0 & 1 & \ddots & & \vdots \\ \vdots & & 1 & 0 & 0 \\ 0 & \cdots & 0 & 1 & 1 \end{bmatrix} \quad B = \begin{bmatrix} 0 & 1 & \cdots & 1 \\ 0 & -1 & & 0 \\ \vdots & & \ddots & \vdots \\ \vdots & & & 0 \\ 0 & 0 \cdots & 0 & -1 \end{bmatrix}$$

The problem facing the profit maximizing timber firm is specified as

$$\max_{c(t)} \pi = \sum_{t=1}^{T} p_t c'(t)\, G \tag{6.14}$$

subject to

$$\left[\begin{array}{cccccc|cccc} I & 0 & 0 \ldots & & & 0 & 0 & & \ldots & 0 \\ -A & I & 0 & & & 0 & -B & 0 \ldots & & 0 \\ 0 & -A & I \ldots & & & 0 & 0 & -B & & \\ \vdots & & & & & \vdots & \vdots & & & \vdots \\ & & & & & 0 & & & & 0 \\ 0 & & & \ldots 0 & -A & I & 0 & & \ldots 0 & -B \end{array}\right] \begin{bmatrix} x(0) \\ \cdot \\ x(T) \\ \hline c(1) \\ \vdots \\ c(T) \end{bmatrix} = \begin{bmatrix} \bar{x}(0) \\ 0 \\ \hline \vdots \\ \\ 0 \end{bmatrix} \tag{6.15}$$

$$x(t),\ c(t) \geqslant 0$$

In (6.14) the growth function is specified as a vector $G' = (0, g_1, \ldots, g_n)$, where $g_i > 0$ is the number of cubic metres per hectare in a stand of age i.[2] Hence $c'(t)\, G = c_{t1} g_1 + \ldots + c_{tn} g_n = c_t$ is the total supply (m^3) of roundwood in period t. $p_t \geqslant 0$ is the present value of the constant profit per cubic metre in period t.

If discounted net prices are independent of the cut, then the profit function is linear in $c'G$ and the restrictions are linear. In other words, the problem is a linear programming problem. One could also have stumpage values depend on the age class. This is done by defining the row vector $p(t) = (0, p_{t1}, \ldots, p_{tn})$, where p_{ti} is the present value of the constant profit per cubic metre of age class i in period t. However, nothing essential is lost by using the profit function (6.14).

[2] The following notational conventions for vectors are used: $p \gg 0$ means that $p_j > 0$ all j; $p > 0$ means that $p_j \geqslant 0$ all j, and $p_j > 0$ for at least one j; and $p \geqslant 0$ means $p_j \geqslant 0$ all j. Moreover, if G is an $(n+1 \times 1)$ vector, then the transpose of G, denoted by G', is a $(1 \times n+1)$ vector obtained by interchanging the rows with the column in G.

Every linear programming problem has its dual. The dual problem to the primal problem (6.14)–(6.15) is written as

$$\min_{\lambda(0)} \nu = \lambda(0)\,\bar{x}(0) \tag{6.16}$$

subject to

$$\begin{aligned} &(a)\quad \lambda(t-1)\,I-\lambda(t)\,A \geqslant 0 && t=1,\ldots,T-1 \\ &(b)\quad -\lambda(t)\,B \geqslant p_t G && t=1,\ldots,T \\ &(c)\quad \lambda(t) \geqslant 0 && t=0,\ldots,T \end{aligned} \tag{6.17}$$

where

$$\lambda(t) = (\lambda_{t0}, \lambda_{t1}, \ldots, \lambda_{tn})$$

The dual variables λ_{ti} are the shadow prices (marginal values) of land of age i at time t. The dual problem is to minimize the value of the initial composition of the forest land (minimize the land value) subject to the restrictions that land at time $t-1$ is worth at least as much as land at time t, and that the shadow price of a unit of land at time t is at least as large as the present value of the biomass growing on a particular piece of land, plus the value of seeded land.

Furthermore, from the fundamental dual theorem of linear programming,[3] we know that

Proposition 6.1
The minimum value of the initial endowment of forest land under the restrictions (6.17) is equal to the maximum profit from forest management subject to the resource constraints (6.15).

The reader should note that the minimum value of forest land must equal the competitive price of land. Under perfect competition, the amount people would be willing to pay for forest land does not exceed the maximum profit from forestry. Competition among buyers extinguishes all profits *in excess of the rate of return on investment* (r).[4] This result also illuminates the technique used in

[3] *Duality theorem of linear programming*
For any pair of feasible solutions to the primal and the dual problem, these feasible solutions are optimal if and only if $\nu = \pi$.

[4] Recall that prices are discounted by the ruling interest rate.

section 5.5, where it was shown that the FPO rotation problem can be solved by asking for the maximum annual amount the market is willing to pay for the option to rent land for forestry. Loosely speaking, forest land has value only because of the services in terms of present and future wood supply that are rendered, and hence land value and the present value of the services rendered are equivalent.

6.4 The Solution, and Some of its Properties

To illustrate the main properties of the solution, we solve the dual for $n = T = 3$ which, owing to complementary slackness conditions, means that we know the primal solution. From restriction (6.17c) we know that $\lambda(3) \geqslant 0$, and from (6.17b)

$$\begin{bmatrix} 0 \\ \lambda_{31} \\ \lambda_{32} \\ \lambda_{33} \end{bmatrix} \geqslant \begin{bmatrix} 0 \\ p_3 g_1 + \lambda_{30} \\ p_3 g_2 + \lambda_{30} \\ p_3 g_3 + \lambda_{30} \end{bmatrix} \tag{6.18}$$

Using the finite planning horizon assumption and the fact that no scrap value is assigned to seeded land at the end of the planning horizon, we have $\lambda_{30}^* = 0$. (If we assign a unit of land scrap value S, λ_{30} shall be put equal to S.) Thus, we can write

$$\begin{bmatrix} 0 \\ \lambda_{31} \\ \lambda_{32} \\ \lambda_{33} \end{bmatrix} \geqslant \begin{bmatrix} 0 \\ p_3 g_1 \\ p_3 g_2 \\ p_3 g_3 \end{bmatrix} \tag{6.18a}$$

Moreover, from condition (6.17a), $\lambda(3)A \leqslant \lambda(2)I, \ldots, \lambda(1)A \leqslant \lambda(0)I$, i.e. a unit of forest land at $t-1$ is worth at least as much as the same unit of land at time t. This means that the 'optimal strategy' to minimize v is to choose $\lambda_{31}^* = p_3 g_1$, $\lambda_{32}^* = p_3 g_2$, $\lambda_{33}^* = p_3 g_3$. If $\lambda(3)$ is chosen larger than $\lambda^*(3)$ this might lead to an unnecessarily high value of $\lambda(2)$, and hence indirectly of $\lambda(0)$.

Continuing to solve the problem backwards, for $t = 2$ we obtain

$$\begin{bmatrix} \lambda_{20} \\ \lambda_{21} \\ \lambda_{22} \\ \lambda_{23} \end{bmatrix} \geqslant \begin{bmatrix} \lambda_{31}^* \\ \lambda_{32}^* \\ \lambda_{33}^* \\ \lambda_{33}^* \end{bmatrix} = \begin{bmatrix} p_3 g_1 \\ p_3 g_2 \\ p_3 g_3 \\ p_3 g_3 \end{bmatrix} \tag{6.19}$$

From condition (6.17b) we get

$$\begin{bmatrix} 0 \\ \lambda_{21} \\ \lambda_{22} \\ \lambda_{23} \end{bmatrix} \geqslant \begin{bmatrix} 0 \\ p_2g_1 + \lambda_{20} \\ p_2g_2 + \lambda_{20} \\ p_2g_3 + \lambda_{20} \end{bmatrix} \tag{6.20}$$

i.e. a unit of land at time 2 is worth at least as much as the value of the standing timber plus the value of the newly seeded land. Since we are trying to minimize v, $\lambda_{20}^* = \lambda_{31}^* = p_3g_1$, because $\lambda_{20} = p_3g_1 + \epsilon$ might make λ_{00} at least ϵ too large. Substituting for λ_{20} in (6.20), we obtain

$$\begin{bmatrix} 0 \\ \lambda_{21} \\ \lambda_{22} \\ \lambda_{23} \end{bmatrix} \geqslant \begin{bmatrix} 0 \\ p_2g_1 + p_3g_1 \\ p_2g_2 + p_3g_1 \\ p_2g_3 \quad p_3g_1 \end{bmatrix} \tag{6.21}$$

which, combined with (6.19) enables us to determine λ_{21}^*, λ_{22}^* and λ_{23}^* by choosing[5]

$$\begin{aligned} \lambda_{2i}^* &= \max(\lambda_{3\,i+1}^*, p_2g_i + \lambda_{20}^*) \qquad i = 1, 2 \\ &= \max(p_3g_{i+1}, p_2g_i + p_3g_1) \end{aligned}$$

and we can continue backwards making use of the indicated method. More generally, for $t < T$,

$$\lambda'(t) = \begin{bmatrix} \lambda_{t0} \\ \lambda_{t1} \\ \vdots \\ \lambda_{tn} \end{bmatrix} \geqslant \begin{bmatrix} \lambda_{t+1\,1} \\ \lambda_{t+1\,2} \\ \vdots \\ \lambda_{t+1\,n} \\ \lambda_{t+1\,n} \end{bmatrix} = \lambda(t+1)A \tag{6.22}$$

From the restriction (6.17b),

$$\begin{bmatrix} 0 \\ \lambda_{t1} \\ \lambda_{t2} \\ \vdots \\ \lambda_{tn} \end{bmatrix} \geqslant \begin{bmatrix} 0 \\ p_tg_1 + \lambda_{t0} \\ p_tg_2 + \lambda_{t0} \\ \vdots \\ p_tg_n + \lambda_{t0} \end{bmatrix} \tag{6.23}$$

[5] For $i = 3$, λ_{3i}^* is substituted for $\lambda_{3\,i+1}^*$.

Since we are trying to minimize v, the correct strategy is to choose λ_{ti} as small as possible, subject to the constraints (6.17). The solution is obtained by choosing λ_{ti} equal to the largest of the right hand sides in (6.22) and (6.23). In sum:

Proposition 6.2
The solution of the LP problem (6.16)–(6.17) can be found by using the algorithm:

$$\lambda_{ti} = \max(\lambda_{t+1\,i+1}, p_t g_i + \lambda_{t0})$$

for all t (starting with $t = T$) and all $i < n$.
For $i = n$ the right hand side should be written

$$\max(\lambda_{t+1\,n}, p_t g_n + \lambda_{t0})$$

From the way in which the solution is obtained, part (a) of the following proposition is obviously true:

Proposition 6.3
(a) The shadow prices λ_{ti} are independent of the initial endowments $(\bar{x}_{00} \ldots \bar{x}_{0n})$.
(b) The shadow prices λ_{ti} are non-decreasing functions of $p_1, \ldots, p_T$ and $g_1, \ldots, g_n$.

Part (b) of proposition 6.3 is intuitively true. If prices in the future increase, the value of a unit of forest land today cannot decrease.[6] On the other hand, only prices in the future matter. Yesterday's prices cannot affect the value of a unit of today's forest land. Moreover, a 'better' growth function cannot, *ceteris paribus* decrease the value of a unit of forest land.

Intuition also suggests that the stand may be profitably cut if

$$\lambda_{ti} = p_t g_i + \lambda_{t0} \geqslant \lambda_{t+1\,i+1} \tag{6.24}$$

i.e. if the value of the cut plus the value of the seed on a unit of forest land at time t is worth not less than the shadow price of the same unit of forest land at time $t+1$. If the inequality in (6.24) is reversed, timber should be saved. A more formal argument, making use of the Lagrangian of the forest management problem and the Kuhn-Tucker conditions (the complementary slackness conditions), reveals that intuition is indeed correct, i.e.:

[6] The second part of the claim can be demonstrated by invoking an induction argument.

Proposition 6.4
The cutting rules of the linear forest are the following:

(a) Timber of age class i is *cut* at time t if

$$\lambda_{t0} + p_t g_i > \lambda_{t+1\,i+1}$$

(b) It may be *saved or cut* at time t if

$$\lambda_{t0} + p_t g_i = \lambda_{t+1\,i+1}$$

(c) Timber is *saved* if

$$\lambda_{t0} + p_t g_i < \lambda_{t+1\,i+1}$$

Since the cutting rules do not depend on the composition of the initial endowment:

Corollary 6.1
The optimal management of a forest stand in the linear forest does not depend on the size and composition of the initial endowment of forest land $\bar{x}(0) = (\bar{x}_{00}, \bar{x}_{01}, \ldots, \bar{x}_{0n})$.

In other words, the initial endowment may be split into timber firms of any size and composition without affecting the rotation periods of the individual stands or the aggregate supply of timber. Moreover, given that land carrying trees of age class i at time zero, $\bar{x}_{0i}$, is cut along an optimal programme in periods t, $2t$ and T, a doubling of the initial endowment of land carrying this age class at time zero results in a doubling of the supply of wood from stands of the relevant age class in periods t, $2t$ and T. The cutting policies of all other stands are left unchanged.

One could say that corollary 6.1 is the theoretical justification for the practice, more than 150 years old, of treating the forest management problem at a stand level, which we analysed at length in chapters 4 and 5. Things are more complicated with an explicit harvesting technology and/or restrictions on the composition of the total endowments. If the cutting technology exhibits increasing returns to scale, in the sense that the unit cost of harvesting one cubic metre of wood decreases with the area cut, then it pays to merge cuttings together, and optimal harvesting decisions can no longer be taken stand by stand (see also Löfgren, 1984a).

6.5 The Non-linear Forest and the Supply Function for Roundwood

The cost side of the forest management problem has been treated in a rather unsatisfactory manner in sections 6.2–6.4. This is now

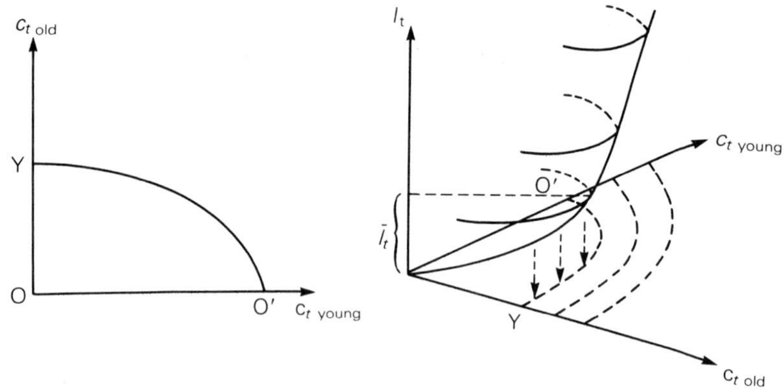

FIGURE 6.1 The harvesting technology: (a) combinations of old and young trees, (b) possibilities resulting from a given labour input

remedied by introducing an explicit cutting technology. To save notational clutter, we assume that labour is the only variable production factor used in the harvesting process.[7] The cutting technology is described by the implicit production function

$$f[l_1, l_2, \ldots, l_T; c(1), \ldots, c(T)] = 0 \tag{6.25}$$

This function defines the vector of required labour inputs given cutting programme $c(1), \ldots, c(T)$. It is reasonable to assume that the technology is separable in the sense that the required labour input in period t depends only on the cut in period t, and not on the cut in other periods. Hence, the function $f(\;) = 0$ may be disaggregated into an implicit function for each period. We write

$$l_t - g_t[c(t)] = 0 \qquad t = 1, \ldots, T \tag{6.26}$$

where $c'(t) = (0, c_{t1}, \ldots, c_{tn})$. To illustrate what is behind the formulation in (6.26), assume that we use $l_t = \bar{l}$ units of labour to cut trees at time t, and that we have two age classes, old and young trees. Equation (6.26) provides the feasible combinations of the areas of land containing old and young trees that can be harvested (see figure 6.1a). We assume that it takes more 'effort' to cut a hectare of old trees than a hectare of young trees. This assumption is reflected by the fact that a given labour input results in a larger area of young trees being harvested than of old, i.e. the distance OO′ is longer than the distance OY in figure 6.1a. Note, however, that the unit

[7] Nothing essential changes if more production factors are introduced.

cost per cubic metre of wood is likely to be higher for the area of young trees, owing to the lower wood content (a smaller harvest in cubic metres).

In figure 6.1b the harvesting technology is depicted in a three dimensional diagram. The harvesting possibilities which result from a given labour input are projected on to the bottom plane. The function (6.26) is assumed to have the properties

$$\text{(a)} \quad \frac{\partial l_t}{\partial c_{ti}} > 0 \qquad i = 1, \ldots, n$$

and (6.27)

$$\text{(b)} \quad \frac{\partial^2 l_t}{\partial c_{ti}^2} \geqslant 0$$

The first property means that a larger input of labour is required to cut a larger area. The second property means that the marginal labour requirement is a non-decreasing function of the area harvested.

We now formulate the forest management problem. Discrimination on the supply side is not necessary, since prices do not discriminate between roundwood of different age classes.[8]

Hence $c_{t1}g_1 + \ldots + c_{tn}g_n = c_t$ is the supply of roundwood at time t. The present value of revenues from forestry may be written as

$$(p_1, \ldots, p_T) \begin{bmatrix} c_1 \\ \vdots \\ c_T \end{bmatrix} = pc$$

To take account of the programme's costs, let

$$w = (w_1, \ldots, w_T)$$

$$l' = (l_1, \ldots, l_T)$$

where w_t is the present value of the wage rate in period t and l_t is labour input in period t. Moreover, define $c^{\mathrm{a}} = [c(1), c(2), \ldots, c(T)]$. *The present value function* is defined as

$$\pi(p, w, \bar{x}_0) = \max_{c^{\mathrm{a}}, l} \{pc - wl: c^{\mathrm{a}}, l \in F \cap C\} \tag{6.28}$$

where

$$F = \{c^{\mathrm{a}}, l: f(l, c^{\mathrm{a}}) = 0\}$$

[8] How to handle different assortments will be dealt with in chapter 9, where we introduce a model containing both pulpwood and sawtimber.

$$C = \{c^{\mathrm{a}}\colon c^{\mathrm{a}} \text{ satisfies } (6.15)\}$$

$$c' = [c(1)\,G, \ldots, c(T)\,G] = c^{\mathrm{a}}G$$

$$\bar{x}_0 = \bar{x}(0) = (\bar{x}_{00}, \ldots, \bar{x}_{0n})$$

The biotechnology C forms a convex and closed set. Moreover, since the amount of land is fixed, it is easily understood that the set of possible harvests is also a bounded set. In other words, the set of possible harvests (supplies) of roundwood is a compact (closed and bounded) and convex set. If F is compact and convex, the feasible set $F \cap C$ will be compact and convex. This fact places certain nice restrictions on the properties of the supply function, i.e. restrictions on the mapping of prices into supplies.[9]

The definition (6.28) states that the maximum present value is a function of the present value price vector p, the present value vector w, and the initial endowment $\bar{x}_0$. The maximum is obtained by choosing the cutting programme c^{a} and the labour input l such that they belong to the feasible set $(F \cap C)$ and provide a maximum value for $pc - wl$. Since this rule specifies how c and l shall be chosen, the maximum profit is no longer a function of the cut and the labour input. The profit maximizing cut becomes a function of the present value price vector p, the wage vector w, and the initial endowment $\bar{x}_0$. This dependence is expressed by the supply and demand vectors (demand for labour)

$$\begin{aligned} c(p, w, \bar{x}_0) &= [c_1(p, w, \bar{x}_0), \ldots, c_T(p, w, \bar{x}_0)] \\ l(p, w, \bar{x}_0) &= [l_1(p, w, \bar{x}_0), \ldots, l_T(p, w, \bar{x}_0)] \end{aligned} \tag{6.29}$$

The present value function in (6.28) embodies all the relevant properties of the technology. The most important of these properties are derived in this section. In chapter 2 we defined the concept of a convex function. Now we show that the present value function is a convex function in p and w, i.e.:

Proposition 6.5 (Convexity)

$$\begin{aligned} \pi(p, w, \bar{x}_0) &= \max_{c^{\mathrm{a}}, l} \{pc - wl\colon c^{\mathrm{a}}, l \in F \cap C\} \\ &= pc(p, w, \bar{x}_0) - wl(p, w, \bar{x}_0) \end{aligned}$$

[9] For formal definitions of the concepts of closed set, bounded set and compact set consult Nikaido (1970). The 'nicest' property of a compact set is that a continuous function takes on both its maximum and its minimum on the set.

is a convex function in p, w, i.e.

(a) $\pi(p^\lambda, w^\lambda) \leqslant (1-\lambda)\,\pi(p', w') + \lambda\pi(p'', w'')$

where

$$p^\lambda = (1-\lambda)\,p' + \lambda p'' \quad 0 \leqslant \lambda \leqslant 1$$

$$w^\lambda = (1-\lambda)\,w' + \lambda w''$$

Convexity means that a weighted average of the present value functions at the two price vectors, p', w' and p'', w'' is at least as large as the present value function at a price vector created by weighting the prices p', w' and p'', w'' by the same weights:

Proof

By the definition of the present value function,

$$p'c^\lambda - w'l^\lambda \leqslant \pi(p', w') = p'c' - w'l'$$

where c^λ, l^λ is profit maximizing for the price vectors p^λ, w^λ, and c', l' is profit maximizing for the vectors p', w'. Similarly,

$$p''c^\lambda - w''l^\lambda \leqslant \pi(p'', w'')$$

For $0 < \lambda < 1$ it will, hence, also be true that

$$[\lambda p' + (1-\lambda)\,p'']\,c^\lambda - [\lambda w' + (1-\lambda)\,w'']\,l^\lambda$$
$$= \pi(p^\lambda, w^\lambda) \leqslant \lambda(p'c' - w'l') + (1-\lambda)\,(p''c'' - w''l'')$$
$$= \lambda\pi(p'w') + (1-\lambda)\,\pi(p'', w'')$$

A convex (concave) function has many important properties. Two of them are that the function is continuous at any interior point and that it is almost everywhere differentiable (see e.g. Katzner, 1970, appendix B).

Assume that the felling programme c^*, l^* solves the maximization problem (6.28), when the net price system is p^*, w^*. By definition it then holds that (note that the vector c is a function of c^a)

$$p^*c^* - w^*l^* \geqslant p^*c - w^*l \quad \text{all } c^a,\ l \in F \cap C \tag{6.30}$$

where $F \cap C$ denotes the feasible choice set. Equation (6.30) says that c^*, l^* makes the present value at least as large as any other feasible felling programme. Multiply the price system by a scalar $\lambda > 0$. Since the restrictions are not changed by the change in the price system and since (6.30) holds, it must also hold that

$$\lambda p^*c^* - \lambda w^*c^* \geqslant \lambda p^*c - \lambda w^*l \quad \text{all } c^a,\ l \in F \cap C \tag{6.31}$$

Hence:

> **Proposition 6.6 (Homogeneity of degree zero)**
> If c^*, l^* are 'profit' maximizing for the price system p^*, w^*, then c^*, l^* are 'profit' maximizing for the net price system λp^*, λw^*, $\lambda > 0$.

We can add a constant cost (T_x), independent of the felling programme, without any change in the present value maximizing cutting programme c^*, l^*, i.e.

$$p^*c^* - w^*l^* + T_x \geqslant p^*c - w^*l + T_x \quad \text{all } c^a,\ l \in F \cap C \qquad (6.32)$$

and hence:

> **Proposition 6.7 (Independence of addition of constant)**
> If c^*, l^* yields a maximum present value for the price system p^*, w^*, it will be present value maximizing if a constant is added to the present value function.

The implications of propositions 6.6 and 6.7 for forest economics are straightforward. If p, w are interpreted as future or expected future net prices, proposition 6.6 says that if all future expected net prices are doubled this will not change timber supply. In other words, timber supply does not depend on the absolute price level. Moreover, one can choose

$$\lambda = \left(\sum_{t=1}^{T} p_t + \sum_{t=1}^{T} w_t \right)^{-1} > 0$$

and obtain the normalized net price system

$$\sum_{t=1}^{T} (p_t' + w_t') = 1, \quad p_t' = \lambda p_t \text{ and } w_t' = \lambda w_t$$

This result should also be compared with the results derived in chapter 4, where it was shown that the Faustmann-Pressler-Ohlin rotation period is homogeneous of degree zero in price and the wage rate (regeneration costs). Proposition 6.7 says that fixed costs (subsidies unconnected with the cut) do not matter for the present value maximizing firm, provided it can stay in business after the fixed costs have been taken care of, i.e. provided that $p^*c^* - w^*l^* + T_x \geqslant 0$.

We now show that the supply of roundwood in a particular period is a non-decreasing function of the discounted net price of roundwood in the same period. The proof contains two steps. First, we prove:

Proposition 6.8 (Supply is neoclassical)
If c^*, l^* are 'profit maximizing' for the price system p^*, w^* and c^0, l^0 are 'profit maximizing' for the net price system p^0, w^0, then

$$(p^* - p^0)(c^* - c^0) - (w^* - w^0)(l^* - l^0) = \Delta p\, \Delta c - \Delta w\, \Delta l \geqslant 0$$

Proof
Since the restrictions (the set $F \cap C$) do not change due to price changes, the following comparisons are possible:

$$p^* c^* - w^* l^* \geqslant p^* c^0 - w^* l^0$$

and

$$p^0 c^0 - w^0 l^0 \geqslant p^0 c^* - w^0 l^*$$

These inequalities can be written as

$$p^*(c^* - c^0) - w^*(l^* - l^0) \geqslant 0$$
$$-p^0(c^* - c^0) + w^0(l^* - l^0) \geqslant 0$$

and by addition one gets

$$(p^* - p^0)(c^* - c^0) - (w^* - w^0)(l^* - l^0) = \Delta p \Delta c - \Delta w \Delta l \geqslant 0$$

Secondly, we use the following corollary:

Corollary 6.2
If only the tth price differs between the vectors p^*, w^* and p^0, w^0, then

$$(p_t^* - p_t^0)\,(c_t^* - c_t^0) = \Delta p_t \Delta c_t \geqslant 0$$

Proof
The corollary follows directly from proposition 6.8.

In other words, a higher net price of roundwood in one period *ceteris paribus* means that at least the same amount of roundwood will be supplied in the period in question. Put differently, the supply curve is not backward bending.

Another interpretation of this result is obtained if discounted prices are thought of as expected net prices. If a forest owner expects

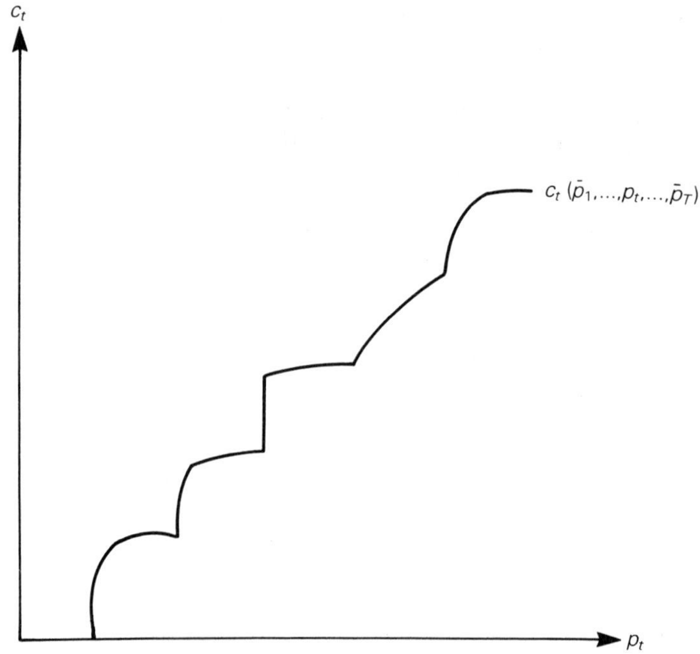

FIGURE 6.2 The supply curve of roundwood

that prices will increase in a particular period *ceteris paribus*, he will supply at least the same amount of roundwood as before the change in price expectations. Moreover, if the only thing that differs between the two price vectors is the wage rate in period t, $w_t^* \neq w_t^0$, from proposition 6.8 $\Delta w_t \Delta l_t \leqslant 0$, i.e. an increased wage rate (cutting cost) decreases the labour input in period t.

The supply curve in period t looks like that in figure 6.2. The supply curve includes kinks to highlight the possibility that the profit maximizing supply is continuous, but non-differentiable at single points.[10] The bar over the price variables $p_j \neq p_t$ indicates that the supply curve is based on the assumption that these prices are kept constant.

[10] The supply curve could be interpreted as composed of a vertical sum of individual forest owners' supply curves. The supply curve could also include vertical segments, i.e. it is a mapping and not a function in the mathematical sense of the word. The mapping is semi-continuous

We now prove the following well-known result, often referred to as Hotelling's lemma, assuming that the present value function is differentiable:

Proposition 6.9 (Hotelling's lemma)

$$\frac{\partial \pi}{\partial p}(p, w, \bar{x}_0) = \pi_p(p, w, \bar{x}_0) = c(p, w, \bar{x}_0)$$

$$\frac{\partial \pi}{\partial w}(p, w, \bar{x}_0) = \pi_w(p, w, \bar{x}_0) = -l(p, w, \bar{x}_0)$$

$$\left.\begin{array}{l} c_t > 0 \\ l_t > 0 \end{array}\right\} \quad \text{all } t$$

This result means that the cutting programme that maximizes present value is obtained by differentiating the present value function with respect to p, and that the present value maximizing labour input programme is obtained by differentiating the present value function with respect to w. In other words, the first order derivatives of the present value function have all the information concerning the supply of roundwood over the planning horizon. Note also that differentiability of the present value function with respect to the price vector means that the supply functions are single valued. The proof of proposition 6.9 is not too difficult:

Proof
Let p', w' be an arbitrary price vector and $c', l' \gg 0$ the corresponding cutting programme that maximizes present value. Now define the function

$$h(p, w, \bar{x}_0) = pc' - wl' - \pi(p, w, \bar{x}_0)$$

This function has, clearly, a maximum for

$$p = p', \ w = w' \, [h(p', w', \bar{x}_0) = 0]$$

Since it is an interior maximum,

$$c'(p', w', \bar{x}_0) - \pi_p(p', w', \bar{x}_0) = 0$$
$$-l'(p', w', \bar{x}_0) - \pi_w(p', w', \bar{x}_0) = 0$$

and since p', w' was arbitrarily chosen, this establishes the claim.

If we assume that the present value function is twice differentiable with respect to p and w, from the convexity of the function in prices and proposition 6.9, the first order derivatives of the supply functions are related in the following way:

Proposition 6.10 (Positive semi-definite quadratic form)

(a) $$\frac{\partial c_t}{\partial p_t} = \frac{\partial}{\partial p_t}(\pi_{p_t}) = \frac{\partial^2 \pi}{\partial p_t^2} \geqslant 0 \qquad t = 1, \ldots, T$$

(The supply curve slopes upwards.)

(b) $$\frac{\partial l_t}{\partial w_t} = -\frac{\partial}{\partial w_t}(\pi_{w_t}) = \frac{\partial^2 \pi}{\partial w_t^2} \leqslant 0 \qquad t = 1, \ldots, T$$

(The demand curve for labour slopes downwards.)

(c) $$\frac{\partial c_t}{\partial p_j} = \frac{\partial}{\partial p_j}(\pi_{p_t}) = \frac{\partial}{\partial p_t}(\pi_{p_j}) = \frac{\partial c_j}{\partial p_t}$$

(The cross-price effects are symmetric.)

(d) The quadratic form $D^2\pi(p, w, \bar{x}_0)$ is positive semi-definite.

Results (a), (b) and (d) follow from the properties of a convex function, and (c) is a consequence of a twice continuously differentiable profit function and Young's theorem for continuous functions $(\partial \pi_{p_t}/\partial p_j = \partial \pi_{p_j}/\partial p_t)$.

These results mean that the supply function is a non-decreasing function of the price, i.e. supply is not backward bending. This has, however, nothing to do with differentiability, since it follows directly from proposition 6.8. The partial derivatives of the supply function are related to each other in an empirically meaningful but fairly complicated manner via the implication of (d).

6.6 The Effects of Taxes

We now use the propositions established in section 6.5 to derive the theoretical effects of taxation on the supply of roundwood. A *proportional tax* on the present value of the harvest in each period is equivalent to a multiplication of the price system p^* by the scalar $\lambda = 1 - t > 0$, where $0 < t < 1$ is the proportional tax rate. Hence, the optimal cutting programme c^* is not changed by the imposition of a proportional tax. This follows directly from the homogeneity property of c^*, l^* (proposition 6.6).

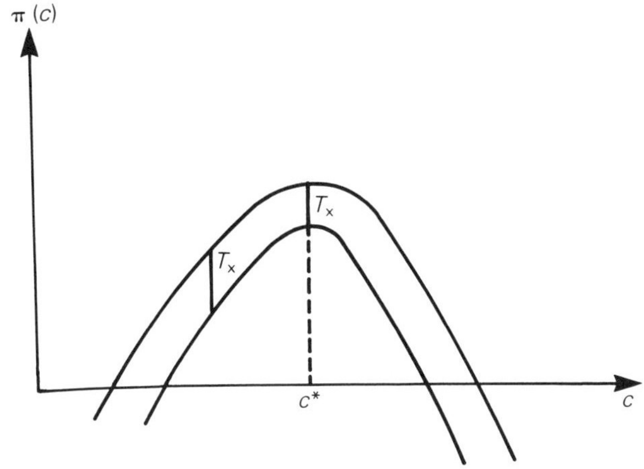

FIGURE 6.3 The effects of a lump sum tax

A *lump sum tax*, the magnitude of which is determined by the site quality classification of the forest land, means that the firm must pay the same amount each period, independent of the cutting programme. Over the entire planning period a constant sum independent of the cutting must be paid (T_x). Proposition 6.7 shows that the optimal cutting programme is independent of the size of the lump sum.

The principles involved can be illuminated by the following diagrams. Figure 6.3 shows that the lump sum tax does not change the optimal cutting programme; it just shifts the profit function $\pi(c)$ downwards.[11] The proportional profit tax does not affect the determination of the optimal cutting programme since it does not influence the relative costs and revenues. This is displayed in figure 6.4. It should also be obvious that a switch from a proportional tax to a lump sum tax does not affect the supply of roundwood.

Despite these comments, there is an important difference between a proportional tax system and a lump sum tax system. Consider a case where the tax rate and the lump sum are allowed to vary between periods. Since

$$T_x = \sum_{t=1}^{T} T_t$$

[11] Note that this is not the present value function $\pi(p, w, \bar{x}_0)$, where c, l are chosen such that $pc - wl$ attains its maximum, implying that maximum profit is no longer a function of the cut and the labour input. The dependence of l is suppressed to enable us to illustrate the point by a two dimensional diagram.

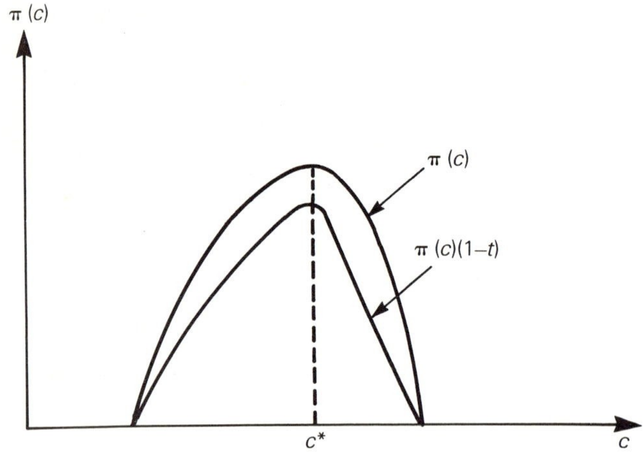

FIGURE 6.4 The effects of a proportional tax

it does not matter if the size of the lump sum varies between periods. Another way to express this is to say that the optimal cutting programme cannot be affected by differences in the level of the lump sum between different periods. If the proportional tax rate is raised in one period, the only difference between the net price vectors before and after the change is the net price in that particular period. From corollary 6.2, the supply of roundwood remains unchanged or decreases. It is thus possible to affect the supply of roundwood by changing the proportional tax rate between periods.

It should also be stressed that expectations of future tax reforms affect the supply of roundwood in the same manner as a future expected price change. For example, expectations about future tax cuts are likely to induce a lower cutting activity today and a higher planned cutting activity tomorrow. A similar effect is created by expectations about future price increases. These effects can be derived from propositions 6.8 and 6.9 and corollary 6.2.

The effect of a progressive tax on the optimal harvesting programme is more complicated since the exact parametrization of the tax system becomes important. However, since the tax rates increase with net revenues, it is intuitively obvious that a progressive tax is likely to even out the harvesting profile over time. The larger the harvest in one period, the lower the marginal net revenue. Thus, it pays to spread out the cut over a longer period. Conversely, a regressive tax is likely to result in a more concentrated harvesting programme.

6.7 The Cost Function and Initial Endowments

Present value maximization requires that the harvesting programme that maximizes present value be obtained at minimum cost, otherwise it would be possible to further increase profits by lowering production costs. Given the harvesting technology (6.25), the cost function is defined as

$$K(c^{\mathrm{a}}; w) = \min_{l} \{wl\colon f(l, c^{\mathrm{a}}) = 0\} = wl(w, c^{\mathrm{a}})$$

and, if the technology has the reasonable separability property defined by (6.26), we can write the cost function as

$$K(c^{\mathrm{a}}; w) = \sum_{t=1}^{T} K_t[c(t); w_t]$$

The cost function yields the minimum cost of cutting when the wage rate is w_t. It enables us to treat the present value maximization problem in a manner somewhat different from the analyses in previous sections. The maximization problem may now be specified as (the wage rate is suppressed to save notational clutter)

$$\max_{c(t)} \sum_{t=1}^{T} \{p(t)\, c'(t)\, G - K_t[c(t)]\} \tag{6.33}$$

subject to the restrictions (6.15).

The Lagrangian of the optimization problem can be written

$$L[c(t), x(t), \lambda(t)] = \sum_{t=1}^{T} \{p(t)\, c'(t)\, G - K_t[c(t)]\}$$

$$+ \sum_{t=1}^{T} \lambda(t)\,[Bc(t) + Ax(t-1) - Ix(t)] + \sum_{i=0}^{n} \lambda_{0i}[\bar{x}_{0i} - x_{0i}] \tag{6.34}$$

with a solution that must fulfil a condition similar to proposition 6.2, i.e.[12]

$$\lambda_{ti} = \max\left(\lambda_{t+1\,i+1}, p_t g_i - \frac{\partial K_t}{\partial c_{ti}} + \lambda_{t0}\right) \qquad \text{all } t \text{ and } i \tag{6.35}$$

[12] The value of the Lagrangian at optimum is a saddle point, i.e.

$$L(c^0, x^0, \lambda) \geqslant L(c^0, x^0, \lambda^0) \geqslant L(c, x, \lambda^0)$$

where c^0, x^0, λ^0 solves the optimization problem.

The earlier strategy of working backwards in determining the shadow prices is no longer feasible since $\partial K_t/\partial c_{ti}$ depends on c_t^a. In other words, the imputed value of a unit of land along an optimal path is no longer independent of the area cut, which in turn cannot be independent of the size and composition of the initial endowment. This also means that timber supply along the optimal programme depends on how the forest land is divided between different present value maximizing forest owners. Say for example, that average cutting cost is a decreasing function of the area cut. If the forest land in this case is divided between many different owners, one would expect a smoother stream of timber compared with a situation where the land is owned by a single owner. The reason for this is that it may pay to merge cuttings when average cost is decreasing. (See the regressive tax case discussed in section 6.6.)

Similarly, if average cutting cost is an increasing function of the area cut, a more concentrated ownership would be expected to result in a cutting profile smoother than the profile under the many owners' case. (See the progressive tax case discussed in section 6.6.) Applying a simple induction argument and using the solution algorithm, one can prove:

Proposition 6.11

Along an optimal programme, assuming

$$\frac{\partial K_t}{\partial c_{tj}} \leqslant \frac{\partial K_t}{\partial c_{ti}} \qquad \text{all } i \leqslant j \text{, and all } t$$

it holds that

$$\lambda_{t0} \leqslant \lambda_{t1}, \ldots, \leqslant \lambda_{tn} \qquad \text{all } t$$

Hence we obtain the reasonable result that the older are the trees, the more valuable is a marginal unit of the forest land. The conditions $\partial K_t/\partial c_{tj} \leqslant \partial K_t/\partial c_{ti}$ for $i \leqslant j$ mean that marginal cutting costs decrease with the age of the stand.

If the cost function is strictly convex, i.e. marginal cutting costs increase with the area cut, the objective function $\pi(c^a)$ is concave in c^a. Similarly, if the cost function is strictly concave in c^a, i.e. marginal cutting costs decrease with the area cut, then the objective function is convex in c^a. It is fairly straightforward to show[13] that the present value function is concave in the initial endowments provided that the objective function is concave in c^a. From the concavity and convexity of the present value function with respect to $\bar{x}_0$ and p, it follows that this function is continuous for $\bar{x}_0 \gg 0$ and $p \gg 0$. Moreover, one can show that it is also differentiable except on sets on Lebesque measure zero (loosely speaking in single points).[14]

Differentiability of the present value function may enable us to use a mathematical result often referred to as the *envelope theorem*. Stated very compactly, according to the envelope theorem

$$\frac{\partial \pi(\theta)}{\partial \theta} = \frac{\partial L}{\partial \theta}(c, \lambda, \theta) \tag{6.36}$$

where θ is a parameter vector containing both p and $\bar{x}_0$, and $L(c, \lambda, \theta)$ is the Lagrange function in (6.34). Thus, if we were to differentiate $\pi(\theta) = \pi(c, \theta)$ with respect to θ, remembering that the optimal c depends on θ, we would have to use the chain rule and thus determine the dependence between c and θ by using the first order conditions for (an interior) maximum. The envelope theorem states that we can get the result indirectly by taking the partial derivative of the Lagrangian with respect to θ, and evaluating it at the optimal c and λ.[15] As the reader has already discovered, the envelope theorem implies Hotelling's lemma (proposition 6.9).

[13] The following lemma yields the desired result.

Lemma
If the objective function $\pi(c)$ – which is independent of $\bar{x}_0$ – is concave, and the constraints define a convex set $C(c, \bar{x}_0)$, then the profit function is concave in $\bar{x}_0$.

Proof
Let x_0' and x_0'' be two parameter vectors and c' and c'' the corresponding optimal programmes. Then (c', x_0') and (c'', x_0'') are in C. Moreover, $\pi(x_0') = \pi(c')$ and $\pi(x_0'') = \pi(c'')$ by definition. For $0 \leqslant \lambda \leqslant 1$ the vector $\lambda x_0' + (1-\lambda)x_0''$, $\lambda c' + (1-\lambda)c''$ is in C since C is convex. Moreover

$$\pi[\lambda x_0' + (1-\lambda)x_0''] \geqslant \pi[\lambda c' + (1-\lambda)c''] \geqslant \lambda\pi(c') + (1-\lambda)\pi(c'')$$
$$= \lambda\pi(x_0') + (1-\lambda)\pi(x_0'')$$

[14] For proof see Katzner (1970), p. 199. The function $y = |x|$ is for example convex, but is not differentiable at $x = 0$. If the function is convex, right hand and left hand derivatives will, however, always exist. The results could therefore be presented in terms of directional derivatives, but these are clumsy and add little.

[15] The proof is not difficult. The reader is referred to Varian (1978), appendix, for a straightforward proof. From the proof it appears that the solution to the optimization problem must fulfil the first order conditions $\partial L/\partial c = 0$ and $\partial L/\partial \lambda = 0$, where the former is most restrictive in the cases to be considered here.

The envelope theorem can be used to analyse the effect of an increased area with trees of age $\bar{x}_{0i}$ on the present value of the forest land. It can be shown that

$$\frac{\partial v^*}{\partial \bar{x}_{0i}} = \frac{\partial \pi}{\partial \bar{x}_{0i}} = \lambda_{0i} > 0 \qquad i = 1, \ldots, n \tag{6.37}$$

The second equality follows directly from the envelope theorem, while the first equality is established by formulating the dual minimization problem corresponding to the non-linear maximization problem. It is, however, no longer true that $\lambda(0)\,\bar{x}_0$, the value of the forest land at optimum valued at the marginal value of the last unit of land, equals the present value of the cut. The objective function of the dual is[16]

$$\begin{aligned} v^*(c^a, \bar{x}_0, \lambda) &= \pi(c^a) - \sum_{t=1}^{T} \sum_{i=1}^{n} c_{ti} \frac{\partial \pi}{\partial c_{ti}} + \sum_{i=0}^{n} \lambda_{0i}\bar{x}_{0i} \\ &= \sum_{t=1}^{T} \left\{ \left(\sum_{i=1}^{n} c_{ti} \frac{\partial K_t}{\partial c_{ti}} \right) - K_t[c(t)] \right\} + \sum_{i=0}^{n} \lambda_{0i}\bar{x}_{0i} \\ &= \sum_{t=1}^{T} \left\{ c(t) \frac{\partial K_t}{\partial c(t)} - K_t[c(t)] \right\} + \lambda(0)\,\bar{x}_0 \end{aligned} \tag{6.38}$$

which is minimized subject to the constraints (6.15). From Euler's theorem of distribution, the first term is negative if the production function exhibits increasing returns to scale (average harvesting costs per unit area fall when the harvesting area increases), zero if there are constant returns to scale, and positive under decreasing returns to scale.[17]

A direct application of the envelope theorem to the Lagrangian of (6.38) yields (6.37). We have shown:

Proposition 6.12
The value of the initial endowments of forest land under the non-linear problem (6.38) valued at the value of the marginal units of land $\lambda(0) = \partial \pi / \partial \bar{x}_0 = \partial v^* / \partial \bar{x}_0$, is smaller (larger) than the present value of the cut under decreasing (increasing) returns to scale. The two values coincide under constant returns to scale.

[16] See Balinski and Baumol (1968) for a more thorough discussion. Formula (6.38) is particular simple as the restrictions are linear in the problem under consideration.

[17] More precisely, the production function is homogeneous of degree >1, $=1$ and <1, respectively. This means that marginal as well as average variable costs will be decreasing, constant and increasing, respectively. For Euler's theorem see e.g. Sydsaeter (1981).

The intuition behind this proposition is that the shadow price of the marginal unit of land underestimates the average value of a unit of land under decreasing returns to scale, and overestimates the average value of land under increasing returns to scale.

The present value function is concave in $\bar{x}_0$ under decreasing returns to scale. Given that it is also twice differentiable, this implies

$$\frac{\partial \lambda_{0i}}{\partial \bar{x}_{0i}} \leqslant 0 \qquad \bar{x}_{0i} > 0 \qquad i = 0, \ldots, n \tag{6.39}$$

i.e. the shadow price of a unit of land decreases when the initial endowment of this land is increased.

The analysis suggests how the cutting profile is changed if non-linearities, due to a non-linear cost function, are introduced. Potential social inefficiencies caused by increasing returns to scale under a spread ownership structure of the forest land are worth further theoretical and empirical investigations, especially since the trade in forest land is heavily regulated in many countries. A marginal unit of land is, in general, worth more to a forest owner who can utilize scale effects than to a forest owner who cannot. If all mutually advantageous transactions took place, then, for example, stands which would never be cut under the status quo could profitably be cut. This is not the only efficiency gain from trade. Lower cutting cost per unit means that fewer resources have to be used in forestry and, hence, that forgone production (the opportunity cost) in other sectors of the economy is smaller. Existing trade regulations are therefore likely to turn existing short run inefficiencies into long run inefficiencies.

Finally, one can ask what happens to the results derived in sections 6.3–6.5 if further restrictions are added to the restriction system in (6.15). Examples include restrictions on the age composition of the forest, restrictions on the amount cut, and linear combinations of the two. The answer is that almost all of the results remain valid. For example, corollary 6.2 on the upward sloping supply curve remains valid even if the restriction system (6.15) is augmented by restrictions on the amounts cut (see e.g. Johansson and Löfgren, 1983b).

Thus the derived results have a large degree of generality. There are, however, important exceptions. If, for example, the forest owner or the forest firm maximizes the present value from forestry subject to the usual biological and technological restrictions and, in addition, has some restrictions on the net income from forestry during specific periods, the supply curve will be no longer upward sloping but backward bending during the periods to which the binding income

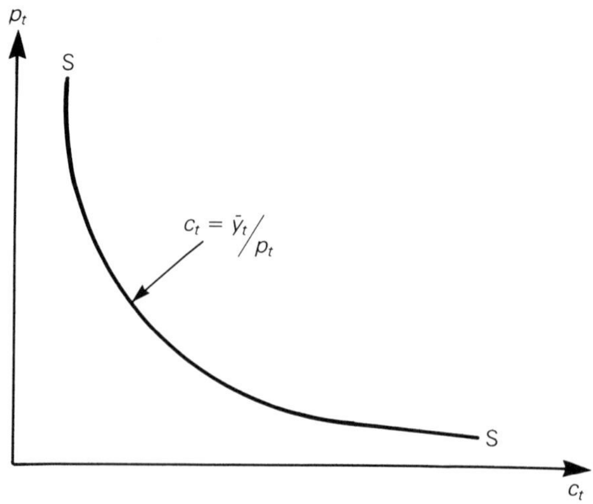

FIGURE 6.5 The backward bending supply curve

constraints refer. If the income restriction (the harvesting cost $w_t l_t$ is disregarded to preserve simplicity) is

$$p_t c_t \geqslant \bar{y}_t$$

where $\bar{y}_t$ is the restriction on the income from forestry in period t, and the restriction is binding, then the supply curve becomes the hyperbola

$$c_t = \frac{\bar{y}_t}{p_t}$$

which clearly is backward bending (see figure 6.5).

The mathematical reason why the proofs in proposition 6.8 and corollary 6.1 are no longer adequate under income restrictions is that the new set of restrictions depends on prices. Hence, the maximization of the present value under the price system p^* is different from the maximization problem under the price system p^0; the feasible sets are different, and derivations in the proof of proposition 6.8 are no longer valid.

Are income constraints realistic? Some people would be inclined to answer in the affirmative. Empirical data on so-called combined forestry/agricultural firms reveal that the owner often uses his forest as a 'bank' ('the Volvo argument') by cutting trees sufficient to finance investments in agriculture.[18] The reason for this seemingly irrational behaviour could be the fact that the capital market is imperfect.

In the Nordic countries where, for example, a large share (more than 45 per cent) of the forest land is owned by private owners whose main source of income is outside forestry, one would expect to discover some signs of income constrained behaviour. To the best of our knowledge, we do not know of any econometric investigation that supports the contention that the supply curve is backward bending. On the other hand, the problem of the combined forestry/agricultural firm deserves further analysis, and this is one of the purposes of chapter 7.

[18] See e.g. von Malmborg (1969). In fact, in Sweden one talks about 'the Volvo argument' for cutting, i.e. the forest owner cuts his timber when he needs a new Volvo. The better are timber prices, the smaller is the necessary cut.

7

The Behaviour of the Self-employed Forest Farmer, and the Forest Industry as a Forest Owner

7.1 Introduction

Chapters 4, 5 and 6 analyse the supply behaviour of any forest owner in a competitive environment. If all forest owners face competitive conditions, there is little reason to treat the behaviour of certain types of forest owner separately. Nevertheless, in this chapter is analysed the behaviour of two different types of forest owner – the self-employed forest farmer and the forest industry as a forest owner. The reasons for this division are the following. In the analysis of the self-employed forest farmer, no explicit assumption of whether he faces competitive conditions or works in imperfect markets is needed. The allocation of time between working in forestry, agriculture or industry is treated explicitly. This enables us to treat the effects on the allocation of his leisure time among different activities caused by different types of taxes and a switch from one system of taxation to another. One reason why the results differ from the results of previous chapters is that a broader spectrum of relative prices, generated from both inside and outside the forest sector, are relevant for the behaviour of the self-employed forest farmer.

The analysis of the forest industry as a forest owner is interesting because it provides us with the properties of the demand function for roundwood, which we need to complete our partial equilibrium analysis of the roundwood market. It also provides new insights into management of the forest industry's own forests under perfect and imperfect conditions in the roundwood market.

The analytical technique used in this chapter is similar to the techniques used in the previous chapters. The behaviour of the self-employed forest owner is treated in sections 7.2, 7.3 and 7.4. Sections 7.5 and 7.6 are devoted to an analysis of the forest industry.

7.2 The Modelling of the Behaviour of the Self-employed Forest Farmer

Assume that the self-employed forest farmer has the utility function

$$U = U(y, \bar{L} - l_1 - l_2 - l_3) \tag{7.1}$$

which is strictly quasi-concave, twice continuously differentiable, and increasing in each argument ($\partial U/\partial y > 0$, $-\partial U/\partial l_i > 0$). The variables have the following meanings:

$\bar{L}$	total available number of hours
l_1	total number of hours supplied for work in forestry
l_2	total number of hours supplied for work in industry
l_3	total number of hours supplied for work in agriculture
y	real consumption of consumer goods
$\bar{L} - l_1 - l_2 - l_3$	leisure time

The assumption that the function is increasing in each argument means that additional goods and/or leisure increase the agent's utility. (Strict) quasi-concavity means that the set of leisure and goods combinations that are considered better than, or at least as good as, the leisure and goods combinations yielding a certain arbitrarily chosen utility level, is (strictly) convex. This restriction is imposed to guarantee an interior maximum.

The fact that $L - l_1 - l_2 - l_3$ is included as one argument in the utility function, instead of l_1, l_2, l_3 as separate arguments, means that the disutility of one hour of work is independent of where the self-employed farmer works. In other words, the farmer has no preference for any particular working place. This assumption simplifies the analysis considerably.[1]

The utility function (7.1) is maximized subject to a budget constraint

$$\pi_f(l_1; p_r) + \pi_a(l_3; p_a) + wl_2 - p_c y \geqslant 0 \tag{7.2}$$

where the variables have the following meanings:

$\pi_f(l_1; p_r)$	(imputed) income from l_1 hours of work in forestry
$\pi_a(l_3; p_a)$	(imputed) income from l_3 hours of work in agriculture

[1] It must be confessed that the assumption is an excellent trick to obtain clear cut results.

w	wage rate in industry
p_c	price of consumer goods
p_r	parameter denoting the influence of the price of roundwood on the imputed income from forestry
p_a	parameter denoting the influence of the price of agricultural products on the (imputed) income from agriculture

The budget restriction simply says that the value of consumption cannot exceed the 'income' of the self-employed forest farmer. The imputed incomes π_a and π_f in a competitive environment may be interpreted as the annual interest on the capital stocks of the agricultural and forest farm respectively. There are, however, problems with this interpretation since this analysis is conducted in a static setting, while annual interest income implies an intertemporal setting. The budget constraint (7.2) is not always appropriate for an intertemporal problem. However, as was shown in chapter 4, under certain conditions it may be optimal to consume the (interest) income from forestry (rK). The reader is referred back to proposition 4.3.

We assume that:

Assumption 7.1

$$\pi_f(0; p_r) \geqslant 0$$

$$\pi_a(0; p_a) \geqslant 0$$

and that:

Assumption 7.2

(a) $$\frac{\partial \pi_f}{\partial l_1} > 0, \quad \frac{\partial \pi_a}{\partial l_3} > 0$$

(b) $$\frac{\partial^2 \pi_f}{\partial l_1^2} < 0, \quad \frac{\partial^2 \pi_a}{\partial l_3^2} < 0$$

This means that the incomes from forestry and agriculture are non-negative (assumption 7.1) and that they are differentiable and increasing functions of the input of labour (assumption 7.2a). Moreover, there are diminishing returns to the inputs of labour (assumption 7.2b) (the profit function is strictly concave in l). A possible shape of the $\pi_f(l_1; p_r)$ function is depicted in figure 7.1.

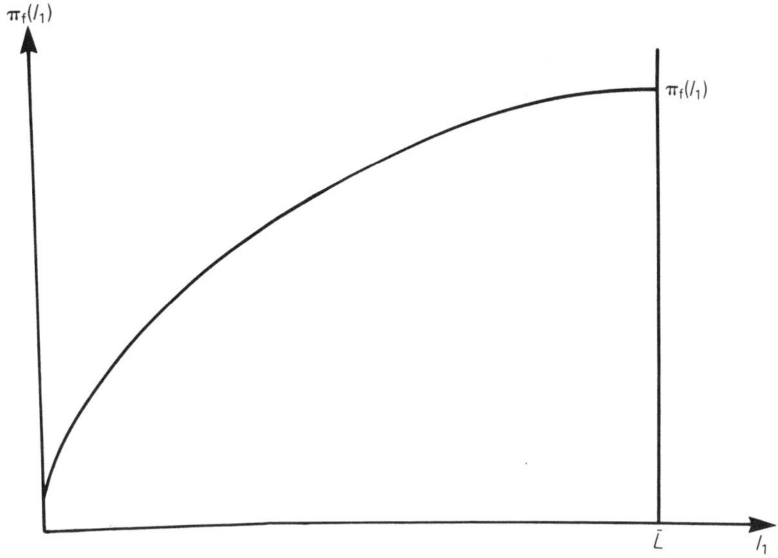

FIGURE 7.1 The capital income from forestry

With respect to the parameters p_a, p_r we assume that:

Assumption 7.3

$$\frac{\partial \pi_a}{\partial p_a} > 0, \quad \frac{\partial \pi_f}{\partial p_r} > 0$$

and that:

Assumption 7.4

$$\left.\begin{aligned} \pi_a(l_3; \lambda p_a) &= \lambda \pi_a(l_3; p_a) \\ \pi_f(l_1; \lambda p_r) &= \lambda \pi_f(l_1; p_r) \end{aligned}\right. \qquad \lambda > 0$$

In other words, the imputed incomes are increasing (assumption 7.3) and linear homogeneous (assumption 7.4) functions of the parameters p_a and p_r. The latter assumption means that an increase in one parameter reflects a uniform increase of all future prices. This implies that the present value of all future incomes increases by factor λ. Thus, the value of the capital (the sum of all future incomes) increases by λ. If the interest rate earned on capital remains constant, this means that the imputed income from capital increases by the same factor.[2] More explicitly, from chapter 4 and the Faustmann-Pressler-Ohlin theorem, we know that the land value of a forest stand (the value of the capital) is given by

$$\max_T \frac{pf(T)\,\mathrm{e}^{-rT}}{1-\mathrm{e}^{-rT}} = \frac{pf(T^0)}{\mathrm{e}^{rT^0}-1} = K$$

If T^0 solves this problem, then from the land rent formulation of the rotation problem in section 5.5,

$$R = \frac{rpf(T^0)}{\mathrm{e}^{rT^0}-1} = rK^0 = \pi_\mathrm{f}$$

Thus, R is the imputed income from forestry and p_r can be interpreted as the wood price p. Moreover, the net revenues from forestry can be written as a function of the input of labour if, for example, thinning, cutting and planting activities are considered more explicitly (see section 6.5).

The self-employed forest farmer's problem is to maximize the utility function (7.1) subject to the budget constraint (7.2). The Lagrange function is

$$L(y, l_1, l_2, l_3, \nu) = U(y, \bar{L}-l_1-l_2-l_3) + \nu(\pi_\mathrm{f}+\pi_\mathrm{a}+wl_2-p_\mathrm{c}y) \tag{7.3}$$

and the first order conditions for an interior maximum are

$$\begin{aligned}
&(\mathrm{a}) \quad U_y - \nu p_\mathrm{c} = 0\\
&(\mathrm{b}) \quad -U_{\bar{L}} + \nu\pi_\mathrm{f}'(l_1; p_\mathrm{r}) = 0\\
&(\mathrm{c}) \quad -U_{\bar{L}} + \nu w = 0\\
&(\mathrm{d}) \quad -U_{\bar{L}} + \nu\pi_\mathrm{a}'(l_3; p_\mathrm{a}) = 0\\
&(\mathrm{e}) \quad \pi_\mathrm{f}(l_1; p_\mathrm{r}) + \pi_\mathrm{a}(l_3; p_\mathrm{a}) + wl_2 - p_\mathrm{c}y = 0
\end{aligned} \tag{7.4}$$

where the following apply:

$\partial U/\partial y = U_y$	marginal utility of consumer goods
$-\partial U/\partial l_i = U_{\bar{L}}$	marginal utility of leisure
$\pi_\mathrm{a}'(l_3; p_\mathrm{a}) = \partial\pi_\mathrm{a}/\partial l_3$, $\pi_\mathrm{f}'(l_1; p_\mathrm{r}) = \partial\pi_\mathrm{f}/\partial l_1$	increase in imputed income of the last hour spent in agriculture and forestry, respectively
ν	Lagrange multiplier

[2] If $\lambda < 1$, the word 'decrease' must be substituted for 'increase'.

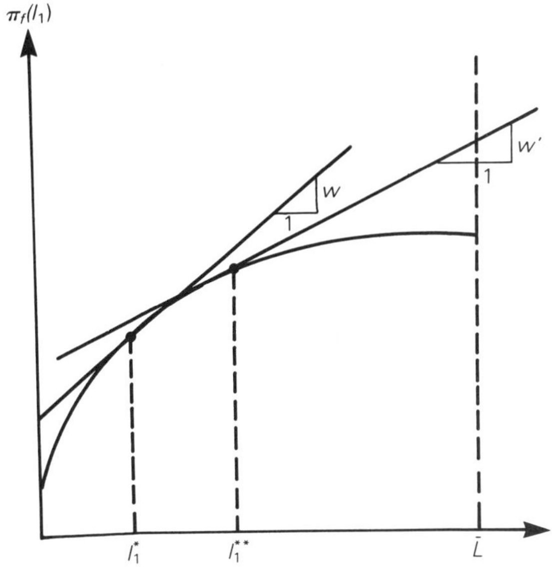

FIGURE 7.2 The optimal self-activity in forestry

There are five equations in five variables (y, l_1, l_2, l_3 and v), and it is easy to solve for l_1 and l_3. Combining (7.4b) and (7.4c),

$$\pi_f'(l_1; p_r) = w \tag{7.5}$$

This equality can be interpreted as an equality between the net return from the last hour's work in forestry and industry, respectively. Since (7.5) contains only one unknown,

$$l_1 = \pi_f'^{-1}(w, p_r) \tag{7.6}$$

Equation (7.6) says that the level of forestry activity depends on the wage rate and the price of wood. An analogous procedure enables us to derive a similar relation for agriculture:

$$l_3 = \pi_a'^{-1}(w, p_a) \tag{7.7}$$

From the equilibrium condition (7.5), we see that the optimal degree of self-activity in forestry can be found diagrammatically, where the slope of the income function coincides with the wage rate (see figure 7.2). The optimal self-activity, given w, is equal to l_1^*. If the nominal wage rate is decreased to w', the new equilibrium is given by $l_1^{**} > l_1^*$. The reason is simple. If the wage rate decreases, the income from the last hour spent working in the industry decreases, while the income from the last hours spent in forestry and agriculture remain unchanged. It is therefore profitable to increase the time spent in agriculture and forestry until the wage rate and the income from the last hours spent in agriculture and forestry are equalized.

Hence:

Proposition 7.1
Given the model of the self-employed farmer specified by (7.1)–(7.3) and assumptions 7.1–7.4, a decrease in the wage rate will increase the time spent in both forestry and agriculture.

It is also easy to show that the level of self-activity is homogeneous of degree zero in parameters p_a, p_r, w, p_c. After the prices are scaled by $\lambda > 0$, the budget constraint can be written

$$\pi_f(l_1; \lambda p_r) + \pi_a(l_3; \lambda p_a) + \lambda w l_2 - \lambda p_c y = 0$$

which, making use of assumption 7.4, can be rewritten as

$$\lambda[\pi_f(l_1; p_r) + \pi_a(l_3; p_a) + w l_2 - p_c y] = \pi_f(l_1; p_r) + \pi_a(l_3; p_a) + w l_2 - p_c y = 0$$

This shows that the feasible choice set is unaffected by an arbitrary positive scaling of the price vector. Hence, the solution to the maximization problem is invariant with respect to the multiplication of the price vector by a scalar larger than zero. Moreover, the degrees of self-activity in forestry and agriculture depend only on the real wage rates w/p_r, w/p_a, as the activity levels are homogeneous of degree zero in prices and each level of labour input depends only on two nominal prices.

In sum:

Proposition 7.2
The solution to the self-employed forest farmer's management problem (7.1)–(7.3) is homogeneous of degree zero in the price parameters (p_a, p_r, p_c, w). In addition, the labour supply in forestry and agriculture depends only on the real wage rates w/p_r and w/p_a (decreasing functions of the real wage rates).

The independence of the number of hours worked in forestry and agriculture and the price of consumer goods p_c is a bit unusual. Every price change has an income and a substitution effect (see any elementary textbook on microeconomic theory). The reason why these effects are no longer important in this context is that the constant wage rate works like a wedge, defining a constant marginal profitability of the last hour worked in forestry and elsewhere. A change in the price of consumer goods will not change the marginal profitability of the last hour worked, and hence has no effect on the number of hours worked in forestry or agriculture. On the other hand, the price change decreases the real price of leisure (w/p_c) and the real value of a given budget. The price change increases the consumption of leisure, i.e. decreases the number of hours worked in the industry (the substitution effect). If it is assumed that leisure is a normal good, the lower real income decreases the consumption of leisure, i.e. increases the number of hours worked in the industry (the income effect). The combined effect is ambiguous.

If we make the not too unreasonable assumption that the supply of timber is a non-decreasing function of the time spent in forestry, i.e.

$$c = F(l_1) \qquad F' \geqslant 0 \tag{7.8}$$

then by substituting (7.6) into (7.8) we get

$$c = F[\pi_f'^{-1}(w, p_r)] \tag{7.9}$$

Finally, by using proposition 7.2, the supply of roundwood can be written as a non-increasing function of the 'real wage rate':

$$c = F\left(\frac{w}{p_r}\right) = F(\underset{-}{w}, \underset{+}{p_r}) \tag{7.10}$$

where the signs under the arguments denote the signs of the first order derivatives. Thus the supply of roundwood does not decrease when the price of roundwood increases. The aggregate supply of roundwood is obtained by a summation of the individual supply functions.

7.3 Income Constraints and the Backward Bending Supply Curve

In this section we investigate the implications of the so-called 'Volvo argument' (the forest owner cuts his timber when he needs a new Volvo), which was mentioned at the end of chapter 6. By adding the restriction that the imputed income from forestry is equal to or larger than $\bar{\pi}$, for the binding equality case

$$\pi_f(l_1; p_r) = \bar{\pi} \tag{7.11}$$

If roundwood prices increase by a scalar $\lambda > 1$, from the homogeneity of the profit function the income constraint can be written as

$$\pi_f(l_1, p_r) = \bar{\pi}\lambda^{-1} \tag{7.12}$$

Since the profit function is an increasing function of l_1, self-activity in forestry is reduced by an increase in the price of roundwood. If the supply of roundwood is an increasing function of l_1, the result of a higher price for roundwood will be a reduced supply from self-employed forest farmers.

The backward bend of the supply curve derived in this manner is due to the fact that the income constraint is formulated only in terms of income from forestry. A more logical way to introduce an income constraint would be to allow income from all sources to meet an income target. As a compromise, introduce an income constraint formulated in the incomes from forestry and work in industry, respectively:

$$wl_2 + \pi_f(l_1; p_r) = \bar{\pi} \tag{7.13}$$

The first order conditions for utility maximum now read (interior solution)

$$\begin{aligned}
&\text{(a)} \quad U_y - \nu p = 0 \\
&\text{(b)} \quad -U_{\bar{L}} + (\theta + \nu)\,\pi_f' = 0 \\
&\text{(c)} \quad -U_{\bar{L}} + (\theta + \nu)\,w = 0 \\
&\text{(d)} \quad -U_{\bar{L}} + \nu\pi_a' = 0 \\
&\text{(e)} \quad \pi_f + \pi_a + wl_2 - p_c y = 0 \\
&\text{(f)} \quad wl_2 + \pi_f = \bar{\pi}
\end{aligned} \tag{7.4'}$$

where θ is the Lagrange multiplier corresponding to the constraint introduced in (7.13), and ν is the Lagrange multiplier corresponding to the proper budget constraint (7.2). Combining (7.4′b) and (7.4′c),

$$\pi_f'(l_1; p_r) = w \tag{7.5}$$

We have already shown that self-activity in forestry under (7.5) is an increasing function of the price of roundwood and, given the assumption with respect to the supply of roundwood, this implies a non-decreasing supply of roundwood.

To sum up, the backward bending supply curve of roundwood seems to be built on the assumption that an income constraint can only be satisfied with income from forestry. As soon as alternative income sources are introduced, the supply curve is no longer backward bending. Intuitively, this is straightforward: the utility maximizing self-employed forest farmer strives to satisfy the income constraint with a minimum loss of leisure time. If the marginal revenue from work in forestry increases, under the restriction (7.13) the forester will increase working time in forestry and decrease working time in industry. However, the institutionalized working hours that prevail in most of industry, at least for males, make the behaviour implied by the so-called 'Volvo argument' not unreasonable, at least when credit markets are tight.

7.4 Taxation and the Supply of Roundwood

In this section a comparison is made between the linear and nonlinear forest model introduced in chapter 6 and the model of the self-active forest farmer introduced in this chapter with respect to the impact on the supply of roundwood of different kinds of taxes. The taxes under consideration here are a lump sum tax, a proportional profits tax, and a progressive profits tax.

Suppose a proportional tax t is levied on all types of income. In the linear and non-linear forest model (LNF) introduced in chapter 6, a proportional tax is equivalent to a multiplication of the net price vector by a scalar $\lambda = 1 - t > 0$ and, according to proposition 6.6, this leaves the optimal cutting programme unchanged. A proportional tax on all types of income in the model of the self-employed forest farmer (SEFF) introduced in this chapter leaves all relative prices constant and, hence, does not change the degree of self-activity in forestry or agriculture.

The same effects occur with a lump sum tax T_x. It does not change any relative prices. Thus, the activities in forestry and agriculture are unchanged, irrespective of which of the two models is valid. To sum up:

Proposition 7.3
A proportional tax[3] on all types of income or a lump sum does not change the optimum in forestry in the LNF model or the SEFF model.

Despite this result, there is an interesting difference between the SEFF and LNF models. Assume that all incomes and profits are

[3] A progressive tax on all kinds of income does change the optimum in the LNF and SEFF models. On the other hand, it is not generally possible to determine exactly how the optimal cutting programme of the LNF model is changed by the progressive profits tax.

taxed by the proportional tax t. Furthermore, let the policy maker change the system of taxation so that the proportional tax on imputed income from forestry and other profits from forestry is replaced by a lump sum tax related to the site quality classification of the land. This does not change the firm's behaviour in the LNF model, but it does increase the marginal profitability of work in forestry in relation to work in the industry in the SEFF model. Hence, the number of hours worked is increased in the SEFF model. Given the supply function (7.9), this increases the supply of roundwood. The reason for this difference is that the after tax wage rate in the industry is of importance in the SEFF model but not in the LNF model.

In Sweden, where quite a few forest owners are self-employed in the sense of the SEFF model, it has been discussed whether a switch from a progressive taxation of income to a lump sum tax would increase the roundwood supply (the propensity to cut has decreased considerably during the late 1970s). One answer is given by the earlier analysis. Note, however, that after the switch it is no longer possible to change the cutting behaviour of the self-active forest farmer within the SEFF model by altering the amount of the lump sum.[4] His behaviour in the market for goods and in the labour market (the number of hours worked in the industry) is, however, changed by the income effects which are created by the change in the magnitude of the lump sum. We return to the effects of lump sum taxes on timber supply in the concluding chapter and consider an LNF model with stochastic future prices.

7.5 The Forest Industry as a Forest Owner

The purpose of this section is to introduce the forest industry as an owner of forest land. We analyse the behaviour of the industrial firm with respect to the management of forest land, the supply of its product and the demand for other production factors. We try to illuminate principles rather than details, and therefore make a couple of simplifying assumptions. Most of the formal results derived in this section hold under more general conditions with or without minor modifications.

We assume that the industrial firm faces a constant price for its product (say pulp) and can buy its production factors at constant

[4] This is a truth with modifications. A sufficiently high lump sum would induce bankruptcy.

prices, i.e. we assume that the firm works in a competitive environment. One realistic interpretation of this assumption is that the firm sells its product in international markets, and the supply of its output has no considerable influence on the market price. It is therefore appropriate to start the analysis by considering the behaviour of a competitive firm, and then modify this analysis by assuming that the firm has one of its inputs (forest land) internally available. The chapter ends with a discussion of how the analysis is modified when some of the simplifying assumptions are relaxed.

7.5.1 *The Competitive Firm*

We assume that the firm produces its output, pulp, using two variable inputs, roundwood and labour. The function relating outputs and inputs is called the production function, and is written

$$q = f(c, l) \tag{7.14}$$

where q is the quantity produced, c is the input of roundwood and l is the input of labour.

To facilitate the analysis and guarantee a global maximum, it is assumed that the function is strictly concave and twice continuously differentiable.

The objective of the firm is to maximize profits π with respect to the use of the inputs c and l:

$$\pi = Pf(c, l) - pc - wl \tag{7.15}$$

P is the competitive price of pulp and p and w are the competitive prices of roundwood and labour, respectively. The first order conditions for an interior maximum can be written

$$\begin{aligned} Pf_c(c, l) - p &= 0 \\ Pf_l(c, l) - w &= 0 \end{aligned} \tag{7.16}$$

where

$$f_c = \frac{\partial f}{\partial c}, \; f_l = \frac{\partial f}{\partial l}$$

The partial derivatives f_c and f_l are the marginal products of roundwood and labour, respectively. The first order conditions mean that the firm uses more of an input as long as the value of its marginal product exceeds its price. At an optimum, the values of the marginal products coincide with the input prices. From our assumption about

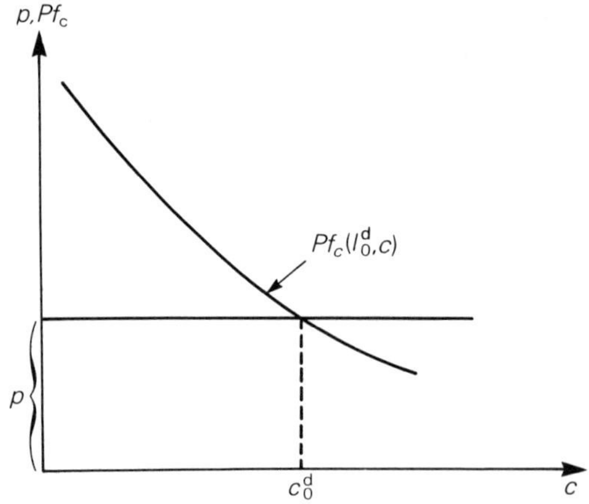

FIGURE 7.3 The demand for roundwood

the production function, it follows that an input combination fulfilling (7.16) constitutes a global optimum.

A partial equilibrium illustration of the determination of the input of roundwood is given in figure 7.3. The firm demands more roundwood as long as the value created by the last unit bought exceeds the price of roundwood.[5] The optimum c_0^d is obtained when the price of roundwood and the value of the marginal product coincide.

If the equations are solved for c and l, one obtains the derived demand functions

$$\begin{aligned} c^d &= c^d(P, p, w) \\ l^d &= l^d(P, p, w) \end{aligned} \tag{7.17}$$

which state that the demand for a production factor is, in general, a function of all factor prices and the price of the product. By substituting (7.17) into the production function, we obtain the supply function

$$q^s = f[c^d(P, p, w),\ l^d(P, p, w)] = q^s(P, p, w) \tag{7.18}$$

Some of the properties of the demand and supply functions are easy to derive. The following proposition holds true:

[5] l_0^d is best thought of as the optimal input of labour, which is in some way known in advance.

Proposition 7.4
The functions in (7.17) and (7.18) are homogeneous of degree zero in prices, i.e.

$$c^{d}(P,p,w)=c^{d}(\lambda P,\lambda p,\lambda w) \qquad \lambda>0$$
$$l^{d}(P,p,w)=l^{d}(\lambda P,\lambda p,\lambda w)$$
$$q^{s}(P,p,w)=q^{s}(\lambda P,\lambda p,\lambda w)$$

In other words, if all prices are multiplied by a scalar $\lambda>0$, factor demands and output are unaffected. If c^*, l^* maximizes profit when prices are P, p, w, then

$$\begin{aligned} Pf_c(c^*,l^*)-p&=\lambda[Pf_c(c^*,l^*)-p]=0\\ Pf_l(c^*,l^*)-w&=\lambda[Pf_l(c^*,l^*)-w]=0 \end{aligned} \tag{7.19}$$

and the proposition is established. It may also be shown that the supply of the product is a non-decreasing function of its price, and that the demand for a production factor is a non-increasing function of its factor price:

Proposition 7.5

(a) $(P''-P')(q''-q')=\Delta P\Delta q\geqslant 0 \qquad \text{for } \Delta p, \Delta w=0$

(b) $(p''-p')(c''-c')=\Delta p\Delta c\leqslant 0 \qquad \text{for } \Delta P, \Delta w=0$

(c) $(w''-w')(l''-l')=\Delta w\Delta l\leqslant 0 \qquad \text{for } \Delta P, \Delta p=0$

The proof is analogous to the proof of proposition 6.7 and is therefore omitted. The reader at this stage should be quite familiar with the 'revealed preference' technique of proof.[6] The results mean that an increased price of pulp in the world market increases the supply of pulp (proposition 7.5a), and that an increased price of roundwood (labour) decreases the demand for roundwood (labour).

Although it is not possible to provide a formal proof, without additional assumptions on the properties of the production function, it is also reasonable to assume that an increased price of pulp increases the demand for both roundwood and labour. The aggregate demand for roundwood is obtained by adding together the demand functions

[6] It could be added that if the supply and demand function are differentiable the inequalities of proposition 7.5a–c will be strict almost everywhere. The reader is invited to derive the counterpart to proposition 6.9 (Hotelling's lemma) by assuming that the profit function $\max_{c,l} \pi$ is differentiable.

of the individual firms in the industry. The aggregate demand function will, of course, share the qualitative properties specified by propositions 7.4 and 7.5.

Before we leave this section, we introduce an alternative method for deriving aggregate roundwood demand. This method is frequently used in empirical analyses of the forest industry. Assume a fixed coefficient production function. This means that in order to produce one unit of pulp one needs a_{cp} units of roundwood and a_{lp} units of labour, irrespective of the amount produced. Assume further that pulp mill capacity is given by $\bar{x}$. The maximum revenue that the pulp mill can obtain is $P\bar{x}$, and the variable costs of production at full capacity utilization are $(pa_{cp} + wa_{lp})\,\bar{x}$, where pa_{cp} is the cost of wood and wa_{lp} is the cost of labour per unit of output. A profit maximizing pulp mill will produce pulp if and only if revenues cover variable production costs. Fixed costs, by definition, are incurred whether or not the firm produces. However, when variable costs are no longer covered, the firm loses more money than fixed costs, and it shuts down the factory.[7]

We now use these findings to derive the aggregate demand curve for roundwood in the pulp and paper industry. A pulp mill with a fixed coefficient technology will stay in business if and only if

$$P \geqslant pa_{cp} + wa_{lp} \tag{7.20}$$

i.e. if the price of pulp covers variable unit costs. If this is the case, the profit maximizing firm will use its full capacity and demand $a_{cp}\bar{x}$ units of roundwood. By raising the price of roundwood, assuming that the prices of labour and the technology are kept constant, the break-even roundwood price p^{b} is reached. If the price is higher than the break-even roundwood price then the factory is closed, but at any price less than or equal to the break-even price, the mill demands $a_{cp}\bar{x}$ units of roundwood (see figure 7.4a). The break-even price

$$p^{\mathrm{b}} = \frac{P - wa_{lp}}{a_{cp}} \tag{7.21}$$

represents the pulp or paper mill's ability to pay for the wood. Assuming there are firms with similar fixed coefficient technologies and different abilities to pay for wood, one can rank these abilities

[7] We have in this analysis neglected fixed costs since, from a social cost–benefit standpoint, they are irrelevant for the decision to stay in business or not. They are always irrelevant to the decision on how much of variable inputs to use. Their levels are, however, not uninteresting to the shareholders even though they are 'sunk' costs from society's point of view.

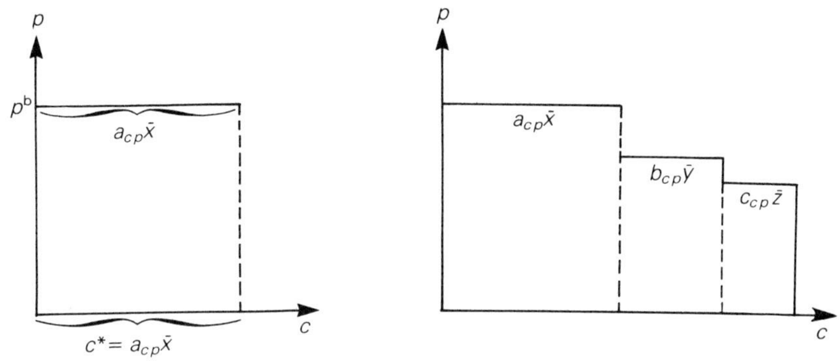

FIGURE 7.4 Demand for roundwood at the level of (a) the firm and (b) the market (aggregate)

in a descending order together with the demand for roundwood at full capacity utilization.[8] This is done in figure 7.4b, where two additional firms are introduced with capacities $\bar{y}$ and $\bar{z}$. The resulting step function constitutes the aggregate demand curve for roundwood.

This approach (see Salter, 1960) is a powerful empirical method that is used in analysing industrial structure. The simplifying assumptions, mainly the constant coefficient technology, do not deviate a lot from reality in the pulp and paper industry.

7.5.2 *The Firms as Forest Owners*

In this section we assume that the firm owns forest land, the biotechnology of which can be described by the LNF forest introduced in chapter 6. The firm now has the option to use roundwood from its own land in the production of pulp or sell some of it in the roundwood market. The crucial question is whether the linear forest will be managed in a manner different from that in Chapter 6. Put differently, assume that the profit maximizing firm owns a given amount of roundwood $\bar{c}$ which it can use as an input in the pulp industry or sell in the roundwood market. How will this amount of roundwood be allocated between different uses? Additional assumptions are needed to answer this question. We need to know wood prices. If roundwood can be bought and sold in a competitive market at a given price, it should not matter whether the firm buys its wood in the roundwood market or uses its own. If it uses a unit of its own

[8] Any other level of capacity utilization can of course be used if there are good reasons to believe that capacity utilization differs from 100 per cent.

wood it loses the revenue from the sales of one unit of wood, and if it chooses to sell the unit, one additional unit of wood has to be bought in the roundwood market.

If the roundwood market is imperfect (in the sense that the selling price of roundwood differs from the price which has to be paid when roundwood is bought in the market), things are clearly different. If the cost price exceeds the selling price, it is reasonable to use all of the wood owned by the firm before anything is bought or supplied to the roundwood market. If the cost price is less than the selling price then the entire stock of wood should be supplied to the roundwood market.[9]

We now show that this intuition is correct. The problem of the firm can be formulated in the following manner:

$$\max_{c,l} \pi = p_1 c_1 + Pf(c_2 + c_3, l) - p_3 c_3 - wl \tag{7.22}$$

subject to

$$\bar{c} - c_1 - c_2 \geqslant 0$$

where the variables have the following meaning:

- c_1 amount of wood supplied to the roundwood market
- p_1 selling price of roundwood
- p_3 cost price of roundwood
- c_3 amount of wood bought by the firm for use in pulp production
- c_2 amount of roundwood from own sources used in pulp production
- $\bar{c}$ available amount of wood owned by the firm

and P, l and w are as before.

The firm's problem is to maximize profit, where profit may include the proceeds from selling wood, subject to the restriction that the sum of the sales of roundwood and the amount of roundwood used in pulp production cannot exceed the total available amount of wood. The reader should note that this formulation does not allow the firm to work as an arbitrager and buy roundwood in one market and sell it in another market. Hence, the case when $p_1 > p_3$ is not reasonably handled by the model.

The Kuhn-Tucker conditions for profit maximum can be written

(a) $p_1 - \lambda \leqslant 0$ $\qquad c_1(p_1 - \lambda) = 0$

[9] The firm should in fact buy roundwood from the market and sell it back again. This arbitrage should in the long run equalize prices. For the moment we will leave the question of why an imperfect price structure can be viable; it will be discussed at some length in chapter 8.

$$
\begin{aligned}
&\text{(b)}\quad Pf_c(c_2+c_3,l)-\lambda \leqslant 0 && c_2(Pf_c-\lambda)=0 \\
&\text{(c)}\quad Pf_c(c_2+c_3,l)-p_3 \leqslant 0 && c_3(Pf_c-p_3)=0 \\
&\text{(d)}\quad Pf_l(c_2+c_3,l)-w \leqslant 0 && l(Pf_l-w)=0 \qquad (7.23) \\
&\text{(e)}\quad \bar{c}-c_1-c_2 \geqslant 0 && \lambda(\bar{c}-c_1-c_2)=0 \\
&\text{(f)}\quad c_1, c_2, c_3, l, \lambda \geqslant 0
\end{aligned}
$$

where λ is the Lagrange multiplier associated with the constraint on the available amount of roundwood. λ gives the value to the firm of an additional unit of roundwood ($\mathrm{d}\pi/\mathrm{d}\bar{c}$).[10]

To begin, we prove the following result:

Proposition 7.6
The optimal value to the firm of an additional unit of roundwood λ lies in the interval $0 < p_1 \leqslant \lambda \leqslant \max(p_1, p_3)$.

Proof
From (7.23a) $\lambda \geqslant p_1 > 0$. Now assume $\lambda > \max(p_1, p_3)$. From (7.23a–c) it then follows that $c_1 = c_2 = 0$, and (7.23) will hold with strict inequality ($\bar{c} > 0$), which implies $\lambda = 0$, a contradiction.

As a direct consequence of proposition 7.6 we have:

Corollary 7.1
If $p_1 \geqslant p_3$ then $\lambda = p_1$.

This means that the value of a unit of roundwood cannot be less than the selling price of roundwood, and it cannot exceed the largest of the selling and cost prices of roundwood. In a perfect market, the cost price and the selling price coincide and are equal to λ. Hence, if the roundwood market is perfect, the firm does not gain anything by using its own roundwood in the production of pulp. It is, however, easy to think of a case where the firm uses all of its available

[10] The proof is straightforward for an interior solution ($c_1, c_2, c_3, l > 0$) since

$$\mathrm{d}\pi = p_1\,\mathrm{d}c_1 + Pf_C\,\mathrm{d}c_2 + \underbrace{(Pf_C - p_3)}_{=0}\,\mathrm{d}c_3 + \underbrace{(Pf_l - w)}_{=0}\,\mathrm{d}l$$

and $\mathrm{d}\bar{c} = \mathrm{d}c_1 + \mathrm{d}c_2$. Hence

$$\frac{\mathrm{d}\pi}{\mathrm{d}\bar{c}} = \frac{p_1\,\mathrm{d}c_1 + Pf_C\,\mathrm{d}c_2}{\mathrm{d}c_1 + \mathrm{d}c_2} = \frac{\lambda(\mathrm{d}c_1 + \mathrm{d}c_2)}{\mathrm{d}c_1 + \mathrm{d}c_2} = \lambda$$

The result also follows from the envelope theorem (see section 6.7).

roundwood in the production of pulp. The following proposition is straightforward:

Proposition 7.7
In an imperfect market with $p_3 > p_1$, then either $\lambda = p_3$ and $c_2 = \bar{c}$ or $\lambda < p_3$, $c_3 = 0$ and $c_1 + c_2 = \bar{c} > 0$.

Proof
From proposition 7.6 one has that $0 < p_1 \leqslant \lambda \leqslant p_3$. If $\lambda = p_3 > p_1$ then $c_1 = 0$ from (7.23a). If $\bar{c} > c_2$ then (7.23e) implies that $\lambda = 0$, a contradiction. If $\lambda < p_3$, $c_3 = 0$ by (7.23c) and $\lambda > 0$ implies $\bar{c} = c_1 + c_2 > 0$.

This proposition means that if the firm's cost price exceeds the selling price, then the firm either uses all of its available roundwood in its own pulp mill and possibly buys additional amounts in the roundwood market, or it buys nothing in the roundwood market and perhaps acts as a net supplier in the roundwood market. It is a trivial exercise to prove that the firm will not use any of its own roundwood if the roundwood market is imperfect in the sense that $p_1 > p_3$.

However, the former imperfection ($p_1 < p_3$) is empirically the most relevant. The cost price structure of roundwood of the typical Swedish pulp mill is shown in figure 7.5. The cheapest timber is

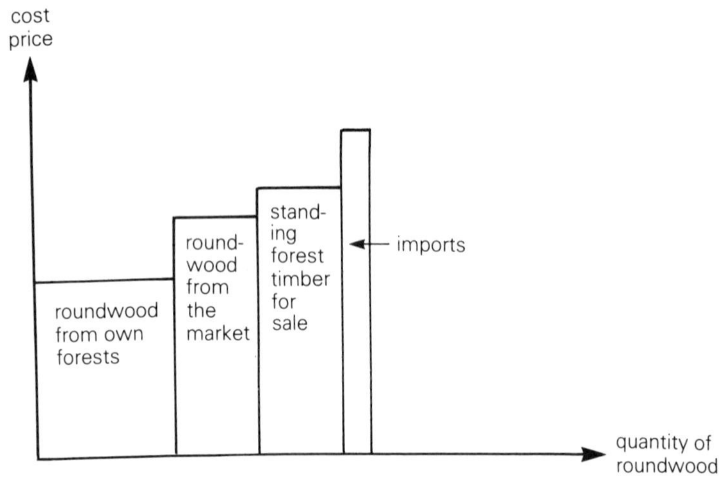

FIGURE 7.5 A typical cost structure for the firm within the Swedish forest industry

obtained from the firm's own forests, market purchases are cheaper than stumpage purchases, and the most expensive way to buy roundwood is in the international market. This cost structure suggests a very imperfect roundwood market and an average selling price for the typical firm which is lower than the cost price of marginal quantities. Hence, it is wrong to say that the cheapest timber is obtained from the own forest since the opportunity cost is at least as high as the market price.

Until now we have assumed that the available amount of roundwood owned by the firm is exogenously given. We have not answered whether the LNF forest will be managed by other principles if it is owned by a pulp producing firm. A satisfactory answer is not difficult to obtain. The total amount of roundwood used in the production of pulp $(c_2 + c_3)$ depends on the price of roundwood, the wage rate and the price of pulp. The demand function has the properties derived in propositions 7.4 and 7.5. It is, however, important to know under what conditions the demand for roundwood is independent of the amount owned by the firm $(\bar{c})$. The following proposition provides an important partial answer:

Proposition 7.8
If $\lambda = p_3 \geqslant p_1$, then the pulp firm's total demand for roundwood will be independent of $\bar{c}$.

Proof
If $\lambda = p_3$, (7.23c) will hold with equality. If (7.23d) holds with equality then $(c_2 + c_3)$ and l, the derived demands, are determined from these equations as functions of P, p_3 and w, i.e. $c = (c_2 + c_3)$ is independent of $\bar{c}$. If (7.23d) holds with inequality, $l = 0$, and c can be determined solely from (7.23c) if l is put equal to zero.

Note that the firm's total profit depends on $\bar{c}$ because, as long as $c_2 + c_3 > \bar{c}$, an additional unit of wood available to the firm would increase profit by $\lambda = p_3$.

It is now time to conclude our discussion with the following proposition:

Proposition 7.9
In a perfect market economy the LNF model will be managed in the same manner, independent of whether it is managed by a pulp firm, or by a firm specialized in forestry, or by private forest owners.

In a perfect market $p_1 = p_3$ and, according to proposition 7.8, the behaviour of the pulp firm will be independent of the amount of roundwood available from internal sources. Hence, all the necessary information to maximize the present value of the forest land and the joint profit of the firm is provided by present and future market prices. This also means that the firm is a net supplier or net demander depending on whether the supply is larger or less than the roundwood demanded by the pulp mill.

The properties of the perfect market are illuminated by noting that there is no reason for the firm to own forest land. The total present value of the firm does not change if the firm sells the forest land at a price equal to the present value generated by the optimal cutting programme, and buys all its roundwood in the roundwood market.[11] The fact that most forest industries own forest land is therefore a clear indication of market imperfections, and also of uncertainty and risk.

How would the optimal cutting policy of the linear forest be changed by an imperfect roundwood market, in the sense that $p_3 > p_1$? From proposition 7.8 the behaviour of the pulp mill is independent of whether roundwood can be supplied internally or not. The value of the linear forest depends on to whom the roundwood is delivered and the opportunity cost of a unit of roundwood, i.e. the maximal contribution to the total profit of the firm of an additional unit of roundwood coincides with the cost price. This means that the optimal cutting policy should be based on today's and future cost cost prices ($\lambda = p_3$) rather than on the market prices.[12]

Finally, it should be stressed that the properties of the supply function of roundwood remain valid even if the forest land is owned by an industry producing final products and using roundwood as an input. On the other hand, a proper modelling of the roundwood market is difficult, not only because there is often a monopsonistic market structure,[13] but also because of prevailing uncertainty about future prices, institutionally conditioned price rigidities, the fact that the forest industry is potentially both a buyer and a seller of roundwood, and the unknown way in which expectations about future prices are formed. Some of these problems will be addressed in the next chapter.

[11] Implicitly this proposition presupposes perfect foresight and a perfect capital market.

[12] This is true as long as the roundwood supplied by the firm from its own forest at cost prices is less than the amount demanded by the firm at those prices. If not, the relevant shadow price lies somewhere in the open interval (p_1, p_3).

[13] One buyer and many sellers.

8

The Roundwood Market under Perfect and Imperfect Market Conditions

8.1 Introduction

In this chapter we conduct a partial equilibrium analysis of the roundwood market using tools developed in chapters 4–7. In addition, in section 8.3 we discuss how a natural resource, in particular the linear forest of chapter 6, fits into a Walras-Arrow-Debreu type of general equilibrium model. We return to the general equilibrium version of an economy with natural resources in subsequent chapters, where we deal with optimal pricing and cost benefit rules for natural resources. In sections 8.4–8.7 we analyse the imperfect roundwood market. Forests have many owners and are spatially rather evenly distributed, while pulp and paper mills and sawmills are often spatially concentrated. Consequently it is likely that few buyers face many sellers of wood. In other words, the roundwood market is likely to be a monopsonistic (or oligopsonistic) market since, by definition, a *monopsony* is a situation where one buyer faces many sellers (and an *oligopsony* is a market situation where few buyers face many sellers).[1] The analysis of the roundwood market as a monopsonistic market structure is conducted in sections 8.4 and 8.5. When the sellers are many and weak and the buyers are few and strong, sellers are likely to organize in some kind of union or association to increase their bargaining strength. Then one seller faces one buyer – a *bilateral monopoly*. In section 8.6 we use a bargaining approach to model some features of the Swedish roundwood market. These features are connected with the price discriminatory approach of sections 8.4 and 8.5.

[1] Compare with the more common terminology of *monopoly* (one seller many buyers) and *oligopoly* (few sellers and many buyers).

8.2 Aggregate Demand and Aggregate Supply

In chapter 7 we derived roundwood demand for a firm using wood as an input in some production process. Demand was shown to be a function of the prices of the final product, wood, and of all other variable production factors. The anaysis was conducted in a static setting, but it is easy to imagine how the demand for inputs in an intertemporal setting would depend on all future prices. Assume that a firm j's demand for wood in a particular period depends only on the price of wood in that particular period, i.e.

$$c_{tj}^{\mathrm{d}} = c_{tj}^{\mathrm{d}}(p_t) \qquad \text{all } t \tag{8.1}$$

For the moment we ignore the influence on wood demand caused by changes in the price of the final products and the prices of other inputs.

The aggregate demand for wood is obtained by summing the individual firms' demand for wood:

$$C_t^{\mathrm{d}} = \sum_{j=1}^{F} c_{tj}^{\mathrm{d}}(p_t) \qquad \text{all } t \tag{8.2}$$

where F is the number of demanders. Equation (8.2) is the aggregate version of (7.17).

The aggregate supply of roundwood is obtained in a similar manner by aggregating the individual agents' supply functions, which are the N forest firms' and/or private forest owners' supply curves derived in chapters 6 and 7:

$$C_t^{\mathrm{s}} = \sum_{h=1}^{N} c_{th}(p) \qquad \text{all } t \tag{8.3}$$

A partial multimarket equilibrium is defined as

$$C_t^{\mathrm{d}}(p_t^*) = C_t^{\mathrm{s}}(p^*) \qquad \text{all } t \tag{8.4}$$

i.e. as the equality between supply and demand in each period. The price vector $p^* = (p_1^*, \ldots, p_t^*, \ldots, p_T^*)$ which creates the equalities specified by (8.4) is called the partial multimarket equilibrium price vector. We assume that an equilibrium price vector exists.[2] In addi-

[2] The existence of equilibrium can be formally proved by assuming continuous excess demand functions, and using the fact that the feasible set is a compact and convex set. See Borek (1976).

tion, by assuming that demand is a strictly decreasing function of the price of wood in the particular period, the uniqueness of the partial multimarket equilibrium of the model specified by (8.1)–(8.4) can be proven:

Proposition 8.1
The partial multimarket equilibrium defined by the model (8.1)–(8.4) is unique provided that

$$\frac{dC_t^d}{dp_t} < 0 \qquad \text{all } t$$

Proof
Assume that p^*, c^* and p^0, c^0 constitute two partial equilibrium vectors. From proposition 6.7 it follows that

(a) $(p^* - p^0)(c^* - c^0) \geqslant 0$

and from the fact that the demand functions are strictly decreasing it follows that

$$(p_t^* - p_t^0)(c_t^* - c_t^0) < 0 \qquad \text{all } t$$

which means that

(b) $(p^* - p^0)(c^* - c^0) < 0$

a contradicton that proves the proposition.

In figure 8.1 we illustrate the unique equilibrium in a one market setting, implicitly assuming that the markets $i \neq t$ are also in equilibrium.

A natural question regarding stability is whether there exist forces that, whenever the price vector differs from p^*, drive prices towards the equilibrium price vector. The usual assumption is to invoke the law of supply and demand and assume that price adjustments are increasing functions of the excess demands, i.e.

$$\dot{p}_t = h(C_t^d - C_t^s) \qquad \text{all } t \tag{8.5}$$

$$h' > 0 \quad \text{and} \quad h(0) = 0$$

$$\dot{p}_t = \frac{dp_t}{dt}$$

In terms of the one market approach of figure 8.1, (8.5) means that if the ruling price is above the equilibrium price (p_t^0) the price will fall towards the equilibrium price. Suppose, for example, that sellers

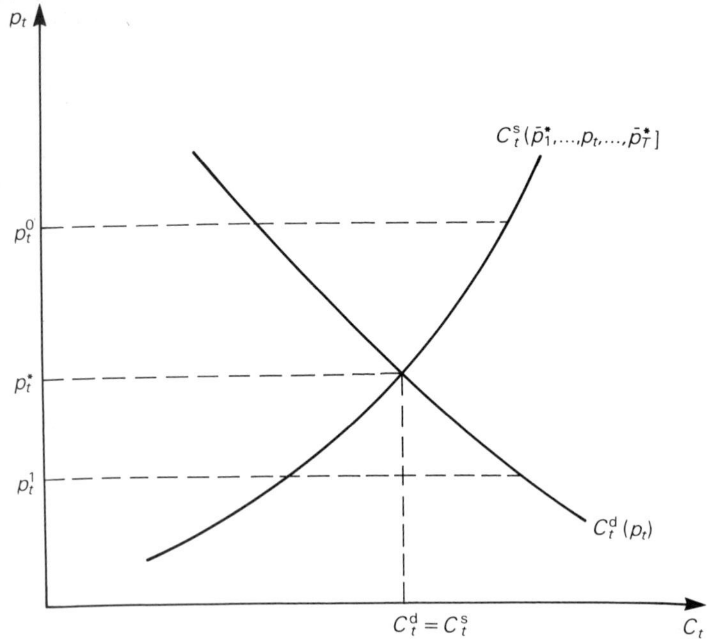

FIGURE 8.1 The equilibrium in market t

underbid each other. If the prevailing price is lower than the equilibrium price (p_t^1) the price will rise towards the equilibrium price.

There are a number of problems connected with (8.5). One is that the dynamic adjustment process may take a substantial time. During disequilibrium the market clearing conditions (8.4) are not satisfied. Rational agents, in forming their expectations about prices and sales, cannot realistically believe that any amount can be traded at the ruling prices. Classical supply and demand functions are derived under the assumption that agents are price takers, and can choose their demands and supplies without perceiving any quantity restrictions.[3]

Secondly, given that everybody is a price taker, who or what changes prices?[4] Thirdly, even if (8.5) is valid, restrictions on the

[3] For the properties of the timber supply function under regulatory constraints see Johansson and Löfgren (1983b).

[4] These problems were pointed out in a celebrated paper by Kenneth Arrow. See Arrow (1959).

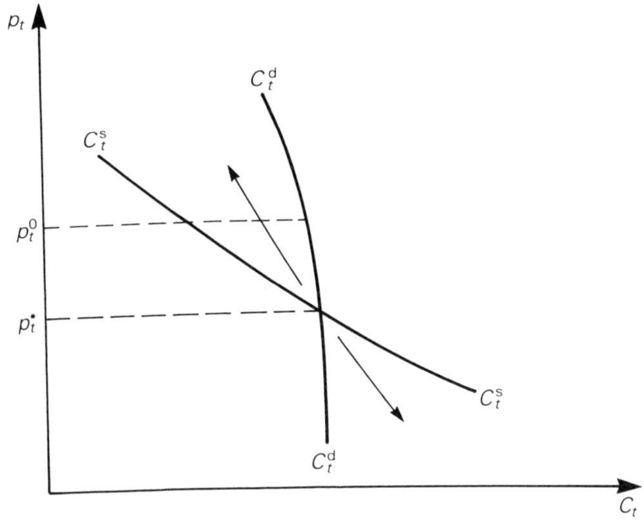

FIGURE 8.2 The unstable roundwood market

properties of the excesss demand functions are needed to ensure that the price adjustment process is stable, i.e. that it converges to the equilibrium price vector p^*. In figure 8.2 we have drawn a downward sloping demand curve and a backward bending supply curve. The negative slope of the demand curve is less steep than the slope of the supply curve, implying an unstable equilibrium. At price p_t^0, excess demand is positive and price starts to rise, thereby generating a larger excess demand, which creates an even stronger upward pressure on prices and an even larger excess demand. The price adjustment process based on (8.5) is unstable and will not close the gap between demand and supply.

The analysis so far has been conducted in a one market setting. Less extreme assumptions are needed to generate an unstable adjustment process in the multimarket case. On the other hand, prices in general show no tendencies to explode and, in modelling price adjustments, one could in most cases assume that the process is fairly stable.

8.3 General Equilibrium: A Few Comments

The linear/non-linear forest presented in chapter 6 constitutes a technology that can be integrated into a general equilibrium model, where other firms and consumers are considered explicitly. More precisely, the model fits fairly well into an Arrow-Debreu world[5] where commodities are dated, i.e. roundwood at date t is treated as being different from roundwood at date $t + 1$. Under competitive conditions, it is possible to prove a number of important theorems on the existence and Pareto efficiency of an equilibrium state. The reader is referred to Arrow and Hahn (1971), where the neoclassical general equilibrium paradigm is rigorously presented.

The properties of a Pareto efficient price system in an economy where natural resources are explicitly considered is discussed at some length in chapter 10. Cost–benefit rules for natural resources are derived in chapter 11. The management of natural resources very often involves externalities which, because of non-zero transactions costs, result in a violation of the perfectly competitive model and may upset the efficiency of an unregulated private ownership economy. Some of these externalities were dealt with in chapters 2 and 3.

8.4 A Monopsonistic Roundwood Market

As mentioned in section 8.1, a resource base like forests often has a large number of owners and is spatially rather evenly distributed. Combined with non-negligible transportation costs and a spatially concentrated wood consuming industry, this is likely to result in a monopsonized (or oligopsonized) market solution. (The local industry gets a spatial monopsony owing to transportation costs.) The purpose of this section is to model the roundwood market as a monopsonistic market structure in a static setting.

The wood buyer is assumed to face a supply price function which is the inverse of the standard supply function:

$$p = p(c) \tag{8.6}$$

where $\partial p/\partial c \geqq 0$, p is the price of roundwood and c is the quantity of roundwood bought by the firm. The supply price of roundwood is a non-decreasing function of the quantity bought.

We further assume that the firm produces its output with one variable input, roundwood. This simplifying assumption is made to save space. Nothing essential is lost. The twice continuously differentiable and strictly concave production function can be written

$$q = f(c) \tag{8.7}$$

[5] See Arrow and Debreu (1954), Debreu (1959) and Arrow and Hahn (1971).

The firm's problem is to maximize profits π:

$$\pi = Pf(c) - p(c)\,c \tag{8.8}$$

where P is the constant[6] (world) market price of the product (say pulp).

The first order conditions for an interior profit maximum can be written

$$Pf_c(c) - \left(p + c\,\frac{\partial p}{\partial c}\right) = 0 \tag{8.9}$$

where

$$f_c = \frac{\partial q}{\partial c}$$

is the marginal product of roundwood. The interpretation of (8.9) is that at the optimal solution, the value of the marginal product of roundwood coincides with the marginal outlay of roundwood.

The situation is illustrated in figure 8.3. The firm's optimum is determined by equating the marginal outlay function and the function representing the marginal revenue product of roundwood. The optimal quantity bought is c^* and this quantity is bought at the market price $p(c^*)$. The competitive market solution is quantity c^0 bought at the price $p(c^0)$, i.e. the quantity bought in the competitive case is determined by equating the factor price and the marginal revenue product of the factor [see (7.16)]. A monopsonistic market has both a lower quantity bought and a lower market price compared with a perfectly competitive market.

By letting $Pf_c = MRP$ (marginal revenue product), one can write the optimum condition (8.9) in the following manner:

$$MRP(c) = p\left(1 + \frac{1}{\eta}\right) \tag{8.9a}$$

where

$$\eta = \frac{\partial c}{\partial p}\frac{p}{c}$$

[6] Nothing essential is changed if one assumes that the firm can affect the market price through its output, i.e. if the firm is assumed to be a member or the sole member of a oligopolistic market structure.

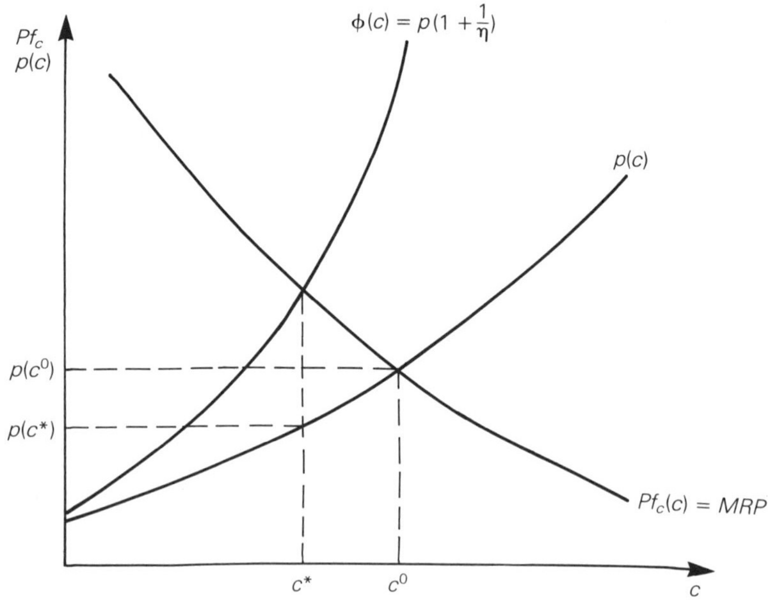

FIGURE 8.3 Equilibrium in the monopsonistic roundwood market

is the supply elasticity in the roundwood market. From (8.9a) it should be clear that a firm perceiving that $\eta = \infty$ acts like a competitive firm.

The demand for roundwood for $\eta < \infty$ is not a function of the price of roundwood. The reason for this is that the firm no longer treats the wood price as a parameter. On the contrary, when $\eta < \infty$, the roundwood price is set by the firm. Solving the first order condition (8.9) for c yields the quantity of roundwood traded and used as an input as an increasing function of the price of the firm's product:

$$c = h(\underset{+}{P}) \tag{8.10}$$

The sign under the argument should be obvious from figure 8.3. An increase in the product price, everything else unchanged, shifts the marginal revenue product function outwards.[7] This results in a larger quantity bought in equilibrium and a higher roundwood price. By

[7] A more formal proof is not difficult. The result will follow by a differentiation of the first order condition (8.9) with respect to P.

substituting (8.10) into (8.6) one obtains the formal relationship between the product price and the wood price:

$$p = p\,[h(\underset{+}{P})] \tag{8.11}$$

The wood price is an increasing function of the price of the final product.

8.5 Price Discrimination under Monopsony

As shown in figure 7.5, a firm (or the forest industry) often buys its roundwood at different price levels, at least in the Swedish forest industry. For this situation to exist, resale must be impossible. One reason why discrimination has continued in the Swedish roundwood market is covered in Löfgren and Ranneby (1984).

Given that the firm can segment the roundwood market into two[8] different non-related submarkets,[8] it faces two different supply price functions

$$\begin{aligned} &p_1 = p_1(c_1) \\ &\qquad\qquad \frac{\partial p_i}{\partial c_i} \geqslant 0 \\ &p_2 = p_2(c_2) \end{aligned} \tag{8.12}$$

where the subscript denotes the market where the roundwood is bought. The objective function of the firm under these circumstances is written[9]

$$\pi = Pf(c) - \sum_{i=1}^{2} p_i(c_i)\, c_i \tag{8.13}$$

where

$$c = c_1 + c_2$$

The roundwood used by the firm is assumed to be homogeneous. Hence we can add the quantities bought in markets 1 and 2 to get an aggregate figure. The first order conditions for a profit maximum can now be written

$$\begin{aligned} Pf_c(c_1 + c_2) - \phi_1(c_1) = 0 \\ Pf_c(c_1 + c_2) - \phi_2(c_2) = 0 \end{aligned} \tag{8.14}$$

[8] As usual this assumption is made to save space. Nothing essential is gained or lost by treating the general case with n submarkets.

[9] Fixed costs are ignored as usual and the objective function is assumed to be twice continuously differentiable and strictly concave.

where

$$\phi_i = p_i \left(1 + \frac{1}{\eta_i}\right)$$

These conditions can be rewritten as

$$Pf_c(c_1 + c_2) = \phi_1(c_1) = \phi_2(c_2) \tag{8.14a}$$

At the optimum, the marginal revenue product of roundwood equals the marginal outlay (cost) in each submarket. The intuitive reason for this optimum condition is that if marginal revenue exceeds marginal costs, it is profitable to buy more wood. Moreover, if the marginal cost is higher in one market than in the other, it is profitable to buy more in the low cost market and less in the high cost market. Redistributions like this are impossible if the first order conditions are fulfilled. Moreover, the roundwood price in the respective submarket is related to the supply elasticity of the supply function at the optimum:

Proposition 8.2
$0 < \eta_1 < \eta_2$ at optimum implies that $0 < p_1 < p_2$.

Proof
The result follows directly from the optimum conditions.

This proposition means that a less elastic supply function at the optimum is associated with lower wood price in this particular submarket. The market situation under price discrimination is illustrated in figure 8.4. In figures 8.4a and b we show the supply functions and marginal outlay functions in the two markets. In figure 8.4c the marginal outlay functions are horizontally summed to form the curve $\Phi(c)$. Moreover, the marginal revenue function $Pf_c(c)$ has been drawn in this figure. The total quantity bought at the optimum is, by construction,[10] determined by the intersection of the marginal revenue product curve and the $\Phi(c)$ curve. The aggregate quantity c^* is then split between the different submarkets by determining the quantities where $Pf_c(c^*)$ equals the marginal outlays in the two markets, i.e. at points A and B in figures 8.4a and 8.4b, respectively. The price level in each submarket is finally obtained by relating quantities to the respective price functions.

[10] This is so because a horizontal summation means that each aggregate quantity of roundwood is bought in such a combination that $\phi_1 = \phi_2$.

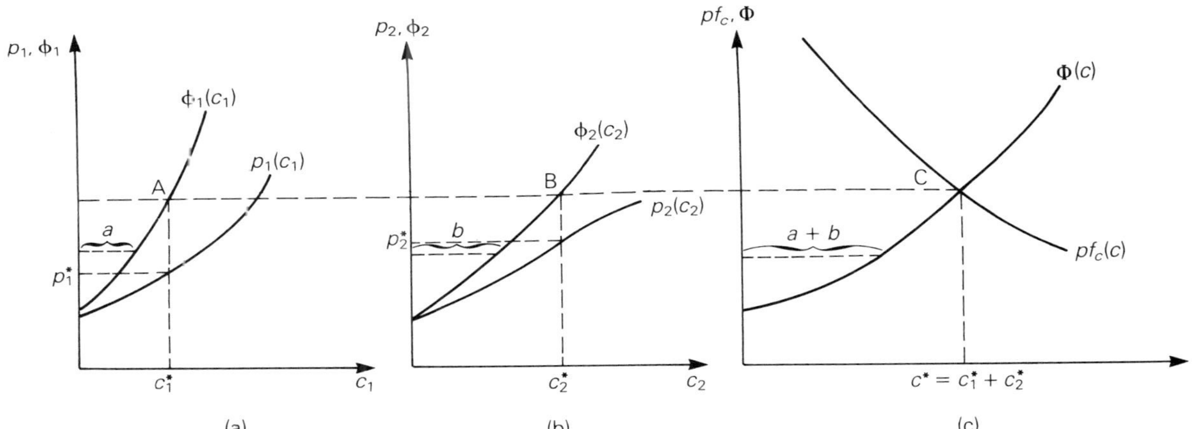

FIGURE 8.4 Price discrimination under monopsony

The properties of the firm's demand functions for roundwood in each submarket are derived by totally differentiating the first order conditions, and using the assumption that the objective function is strictly concave and twice continuously differentiable. The result is that demand, under reasonable assumptions about the slopes of the marginal outlay functions, is an increasing function of the price of the final product,[11] i.e.

$$\begin{aligned} c_1 &= h_1(\underset{+}{P}) \\ c_2 &= h_2(\underset{+}{P}) \end{aligned} \tag{8.15}$$

The reader may easily convince himself that this is the case by checking the comparative statics of a shifting marginal revenue product curve in figure 8.4.

It was shown earlier that monopsony results in a lower quantity bought at a lower price compared with perfect competition. This is also true if price discrimination under monopsony is compared with perfect competition. An important matter is how the total quantity bought under discrimination is related to the quantity which would have been bought had the monopsonistic firm been forced to buy the total quantity at a single price. No simple solution exists, but the following local result contains part of the answer:[12]

Proposition 8.3
Provided that $\eta_2 > \eta_1$ in optimum ($p_2 > p_1$), a marginal transfer from price discrimination towards simple monopsony pricing increases (decreases) the total quantity bought provided that

$$c_2^* \frac{\partial^2 p_2 / \partial c_2^2}{\partial p_2 / \partial c_2} \underset{(<)}{>} c_1^* \frac{\partial^2 p_1 / \partial c_1^2}{\partial p_1 / \partial c_1}$$

For the special case, when the supply functions are linear,

$$\frac{\partial^2 p_2}{\partial c_2^2} = \frac{\partial^2 p_1}{\partial c_1^2} = 0$$

and the two quantities coincide.[13]

[11] What is needed is that the marginal outlay functions both have a positive slope.

[12] The proof is available in Löfgren (1972) and in Ekelund, Higgins and Smithson (1981). For a use of this analytical framework to analyse the Swedish roundwood market the reader is referred to Johansson and Löfgren (1983a).

[13] Note that proposition 8.3 constitutes a local result. For a more comprehensive discussion of global properties the reader is referred to Löfgren (1977) and Formby, Layson and Smith (1983).

8.6 A Bargaining Approach to the Pricing of Wood in Sweden

In a market where buyers have a natural monopsony power, the sellers may organize to counteract the monopsony power of the buyers.[14] This means that the market situation in such a case can be treated as if one seller faces one buyer. Some features of the Swedish roundwood trade fit very well into this pattern, which is termed a bilateral monopoly. The Swedish market is divided into five price regions.[15] Within each region the price levels (the so-called *list prices* of both timber and pulpwood) are set in August every year in negotiations between one organization representing the sellers (the forest owners' association) and one organization representing the buyers. Interregional trade exists, but is more an exception than a rule.

The market characterized by this type of price formation is the proper roundwood market. Buyers, as mentioned in chapter 7, can acquire roundwood in at least three additional ways. They can supply it from their own forests, buy it as standing timber, and import it. The price structure resembles that in figure 7.5.

The wood from their own forests is, as a rule, the cheapest, although its opportunity cost is higher since it can always be sold in the regional or interregional markets subject to transportation cost constraints, as was shown in section 7.4. The list prices in the roundwood market are lower than the cost of standing timber for sale, and the most expensive wood is bought in the international market. One way to explain this price structure is to model the market structure as a monopsony where the buyers are much stronger than the sellers. This is not an unreasonable approach, at least for the period covering the second half of the 1970s and the beginning of the 1980s. During this period the forest industries owned by the forest owners' association (the sellers' side) have had considerable financial problems, and low list prices might have been used as a method (albeit inadequate) to survive. The individual members of the association have realized that they can get better prices for their wood elsewhere, and the forest owners' association has had severe problems supplying wood to its own industries. Many of these industries have either vanished or are owned by other companies. Thus the monopsony price discrimination approach which explains the price structure in the Swedish

[14] This section relies heavily on a paper by Johansson and Löfgren (1984b).

[15] The classification is institutionally determined through an agreement between buyer and seller.

roundwood market during 1975–80, and is covered in section 8.5, has to be modified to cover more general bargaining situations (see also Johansson and Löfgren, 1983a).

This section leans heavily on an approach to wage bargaining developed in McDonald and Solow (1981). By a few modifications of their analysis we are able to cover two separate cases. In the first case the seller is strong (the final price–quantity combination is a point where the marginal revenue product of roundwood is less than the roundwood price), while in the second case the buyer is strong and is able to buy at prices lower than the competitive price of roundwood. This case fits the monopsonistic modelling in sections 8.4 and 8.5, although little emphasis is placed here on the discriminatory pricing market behaviour that is obvious from figure 7.5. Both approaches are primarily aimed at explaining the so-called list prices.[16]

Since the roundwood buyer is likely to have a natural spatial monopoly power, owing to the spatially distributed resource base, we expect the strong buyer to be more likely. On the other hand, during periods of rapid expansion of the capacity of the forest industry, the strong seller might very well apply. Moreover, since the functioning of the roundwood market seems to be imperfect in most countries with a large forest industry, the modelling in this section ought to be of general relevance.

8.6.1 *The Behaviour of the Forest Industry*

Assume that a representative forest company[17] produces a single output with homogeneous roundwood as the only variable input. Moreover, assume that the forest company takes the world market price P of its products as given. The problem of the firm, when no further restrictions are present, is to maximize profits π:

$$\pi = Pf(c) - pc \tag{8.16}$$

The first order condition for profit maximization is

$$Pf_c(c) - p = 0 \tag{8.17}$$

where

$$f_c = \frac{\mathrm{d}f}{\mathrm{d}c}$$

[16] We leave aside the rationing tendencies of the Swedish roundwood market, which were rather frequent during the late 1970s and the early 1980s. Moreover, we do not attempt to explain the decreasing propensity to cut, which has been another important feature of this market.

[17] Best thought of as the forest industry in the particular region.

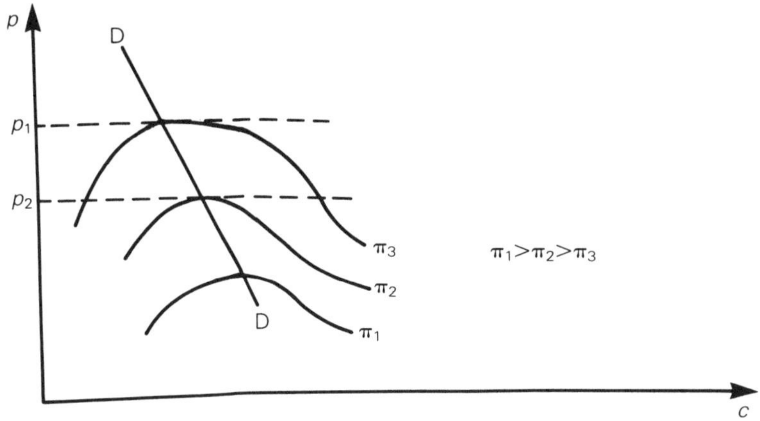

FIGURE 8.5 Isoprofit curves of the forest company

We now use (8.16) and (8.17) to derive *isoprofit curves* giving combinations of p and c that leave profits constant. These curves will later be useful in illustrating the outcome of the 'game'.

Let $\pi = \bar{\pi} =$ constant and take the total differential of the profit function to get[18]

$$d\bar{\pi} = Pf_c(c)\,dc - p\,dc - c\,dp = 0 \tag{8.18}$$

Hence

$$\left.\frac{dp}{dc}\right|_{\pi=\bar{\pi}} = [Pf_c(c) - p]\,c^{-1} \tag{8.19}$$

and

$$\left.\frac{dp}{dc}\right|_{\pi=\bar{\pi}} \gtreqless 0 \quad \text{as} \quad Pf_c(c) - p \gtreqless 0$$

The isoprofit map is shown in figure 8.5. The curves have a positive slope until the price of roundwood reaches the value of the marginal product. Since the production function is strictly concave ($f_{cc}(c) < 0$), the switches of the curves occur at successively lower ps and higher cs. Lower curves represent higher profits; for any given c, a smaller p yields a larger profit. The demand curve for roundwood, DD in the figure, is obtained by selecting combinations satisfying the first order

[18] Expression (8.18) is made equal to zero so it can implicitly represent p and c combinations which generate the same profit level $\bar{\pi}$.

profit maximum condition $Pf_c(c) - p = 0$. In the neighbourhood of the demand curve, the curvature of the isoprofit curves will be as shown in figure 8.5.[19]

8.6.2 *The Behaviour of the Forest Owners*

If forest owners act as if the roundwood market is competitive, the actual price of roundwood equals the supply price. However, the market is not competitive. So we assume that the representative forest owner – or better an association which negotiates roundwood prices – maximizes the utility function

$$F = c[u(p) - \bar{u}] \tag{8.20}$$

where $\bar{u} = u(\bar{p})$ is the (constant) supply price of roundwood, $\bar{p}$ is the (constant) price of roundwood obtained under free competition, p is the actual or negotiated price of roundwood, and c is the amount of roundwood traded.

The difference between $u(p)$ and $\bar{u}$ in (8.20) captures the utility gain per cubic metre roundwood of the 'representative' or average member. This utility gain, multiplied by the total amount c of roundwood sold, denotes the total gain in utility from selling roundwood at prices above the price that the members could obtain under free competition. A two step interpretation of the bargaining process – where price is determined in the first step – which illuminates the particular choice of utility function is given in connection with the derivation of the contract curve in section 8.6.3.

By fixing $c[u(p) - \bar{u}]$ at a certain level, say k, where k is a measure of the bargaining strength of the forest owners or their association, an indifference map may be derived. The *indifference curves* have the usual downward sloping convex shape as in figure 8.6 and converge towards $\bar{p}$ as c increases, provided that the 'strength' parameter k is positive $[u(p) > \bar{u}]$. If k equals zero then the resulting indifference curve is a horizontal line at $p = \bar{p}$ [since $u(\bar{p}) = \bar{u}$]. The case when the strength parameter is negative – yielding indifference curves with positive slopes – is considered in section 8.6.4.

8.6.3 *The Contract Curve when the Buyer is Weak*

In this section, we examine the combinations of roundwood prices and quantities of roundwood which are possible outcomes of efficient

19 $(d^2p/dc^2)_{\pi=k} = \{Pf_{cc}(c)c - 2[Pf_c(c) - p]\}/c^2$

The curves will be concave if $Pf_c(c) > p$ as $f_{cc}(c) < 0$. If $Pf_c(c) < p$ they will be concave if $Pf'_{cc}(c) < 2[Pf_c(c) - p]$.

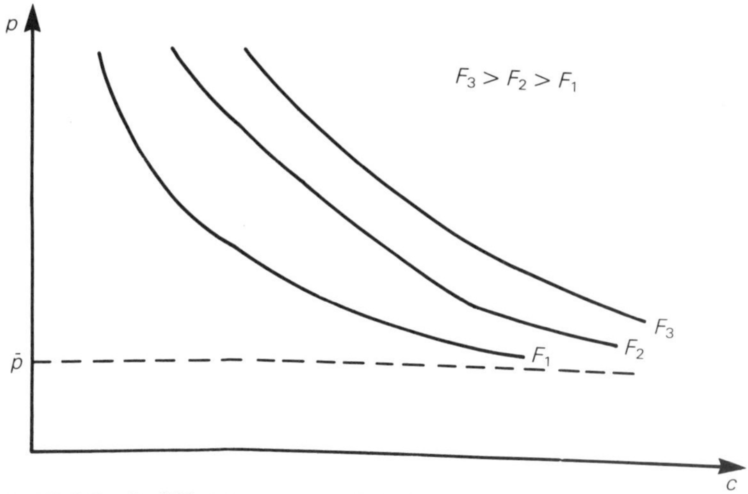

FIGURE 8.6 Indifference curves of the forest owners' association for $k > 0$

bargains. An efficient bargain is a contract which cannot be improved upon by one party without harming (decreasing the utility of) the other party. The reader should convince himself that efficient bargains are the locus of tangency points between the isoprofit curves of the forest company, and the indifference curves of the forest owners' association. We derive the contract curve for $k > 0$, a case where the bargaining power of the association is strong relative to the bargaining power of the industry. Note that $k > 0$ implies that the association receives income and utility over and above the competitive solution.

To obtain the contract curve, the set of efficient bargains, we formulate the following maximization problem:

$$\max_{c,p} \pi = Pf(c) - pc \tag{8.21}$$

subject to

$$c[u(p) - \bar{u}] = k$$

Let the Lagrangian be

$$L = Pf(c) - pc + \lambda\{c[u(p) - \bar{u}] - k\} \tag{8.22}$$

The first order conditions for an interior maximum are

$$\frac{\partial L}{\partial p} = -c + \lambda c u'(p) = 0 \tag{8.23}$$

$$\frac{\partial L}{\partial c} = Pf_c(c) - p + \lambda[u(p) - \bar{u}] = 0 \tag{8.24}$$

$$\frac{\partial L}{\partial \lambda} = c[u(p) - \bar{u}] - k = 0 \tag{8.25}$$

where a prime denotes a partial derivative.

The free market solution is obtained when k equals zero:

$$\begin{aligned} u(\bar{p}) - \bar{u} &= 0 \\ Pf_c(c) - \bar{p} &= 0 \end{aligned} \tag{8.26}$$

i.e. the forest owner is compensated for the marginal disutility of effort and the forest company produces at a point where the value of the marginal product equals the supply price of roundwood.

If $k > 0$ then $u(p) > \bar{u}$ and $Pf_c(c) < p$. The latter inequality means that the firm produces at a point where the value of the marginal product is less than the price of roundwood, i.e. off the demand curve for roundwood.

The bargaining process can be viewed as a two step process rather than one which simultaneously determines both price and quantity. In the first step, the association and the industry bargain on the roundwood price; the association tries to make the average utility gain as large as possible. In the second step, the negotiated price and the exogenously given bargaining strength uniquely determine the volume of wood bought by the industry. This two step view illuminates the possible inefficiencies of a bargaining process. If the first step price bargain is determined without explicit consideration of the relevant efficiency restrictions [given by (8.23) and (8.24) which together define the contract curve], the volume implied by (8.25), given p and k, might be inefficient (off the contract curve). This analysis assumes that the bargains are efficient, which means that price in the first step cannot be determined without considering quantity implications.

To obtain the slope of the contract curve, $\lambda = 1/u'(p)$ is substituted into (8.24). Differentiation after rearranging terms yields

$$\frac{dp}{dc} = Pf_{cc}[u'(p)]^2/[u(p) - \bar{u}]\,u''(p) > 0 \tag{8.27}$$

where $u''(p) = d^2u(p)/dp^2$. The contract curve has a positive slope if $u(p) > \bar{u}$. This is illustrated in figure 8.7. If p equals $\bar{p}$ (the case of free competition) then the 'contract point' is on the demand curve DD of the forest company. As the 'strength' of the association is

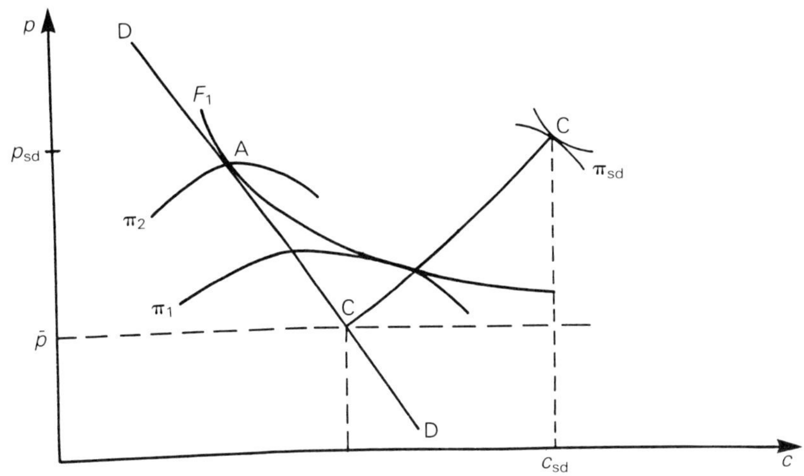

FIGURE 8.7 Isoprofit curves of the forest company (π_1 and π_2), and an indifference curve of the forest owners' association (F_1), the demand curve for roundwood (DD) and the contract curve (CC) when the buyer is weak

increased from zero, the contract curve CC rises to the northeast to end at (p_{sd}, c_{sd}), where the profit of the forest company is so small that it chooses to shut down.

Furthermore, note that a tangency between an indifference curve and the demand curve, like point A in the figure, is Pareto inferior to the relevant points on the contract curve. This can be seen by inspection of figure 8.7. The only exception occurs when the 'strength' of the forest owners equals zero ($k = 0$). Then the contract curve collapses into the demand curve. Lower isoprofit curves represent higher profits, i.e. for a given c a smaller p results in larger profits.

We now investigate the properties of the solution to the bargaining problem, i.e. the properties of p and c as functions of P and k. To do this, eliminate λ from the first order conditions to obtain

$$Pf_c(c) - p + \left[\frac{u(p) - \bar{u}}{u'(p)}\right] = 0$$

$$c[u(p) - \bar{u}] - k = 0. \tag{8.28}$$

These two equations implicitly define c and p as functions of P and k. Totally differentiating (8.28) yields

$$Pf_{cc}(c)\,\mathrm{d}c - \frac{u''(p)[u(p)-\bar{u}']\,\mathrm{d}p}{[u'(p)]^2} = -f_c(c)\,\mathrm{d}P$$

$$[u(p)-\bar{u}]\,\mathrm{d}c + cu'(p)\,\mathrm{d}p = \mathrm{d}k \tag{8.29}$$

Since $u(p)-\bar{u} = k/c$, (8.29) can be written

$$Pf_{cc}(c)\,\mathrm{d}c - \frac{u''(p)}{[u'(p)]^2}\frac{k}{c}\,\mathrm{d}p = -f_c(c)\,\mathrm{d}P$$

$$\frac{k}{c}\,\mathrm{d}c + cu'(p)\,\mathrm{d}p = \mathrm{d}k \tag{8.30}$$

Solving, one obtains

$$\frac{\partial c}{\partial P} = \frac{-f_c cu'(p)}{D} > 0 \qquad \frac{\partial c}{\partial k} = \frac{u''(p)k}{c[u'(p)]^2 D} > 0$$

$$\frac{\partial p}{\partial P} = \frac{kf_c}{cD} < 0 \qquad \frac{\partial p}{\partial k} = \frac{Pf_{cc}}{D} > 0$$

where

$$D = Pf_{cc}cu'(p) + \frac{k^2u''(p)}{c^2[u'(p)]^2} < 0$$

Note that the sign of D is independent of the sign of k. The signs of the partial derivatives are summed up in the following proposition:

Proposition 8.4
The properties of the partial derivatives of the solution to the bargaining problem (8.22) for $k > 0$ are

$$c = c^{\mathrm{w}}(\underset{+}{P}, \underset{+}{k})$$
$$p = p^{\mathrm{w}}(\underset{-}{P}, \underset{+}{k})$$

In other words, an increase in the final product price increases the volume traded, and decreases the roundwood price. This conclusion may appear intuitively unexpected, but can be explained along the following lines. An increased product price induces the buyer to use more wood. Given the strength parameter $k > 0$ a larger volume can, to some extent, compensate for a lower price $\{c[u(p)-\bar{u}] = k > 0\}$. Moreover, an increase of the bargaining power of the association will increase the negotiated price, and increase the quantity traded.

8.6.4 *The Contract Curve when the Buyer is Strong*

In the case of a strong buyer, the model has to be changed. Earlier we assumed that $u(p) > \bar{u}$, so that the actual roundwood price was at least as high as that obtained under free competition $[\bar{u} = \bar{u}(\bar{p})]$. Now reverse this assumption by requiring $u(p) < \bar{u}$. The interpretation of this inequality is that the forest owners, for one reason or another, are unable to obtain the competitive price. The only thing here that differs from the problem which was formulated in (8.22) is that $k < 0$. and this has implications for the slopes of the indifference curves, which resemble those in figure 8.8. The curves converge towards $\bar{p}$ as c is increased, and higher curves represent higher levels of utility. In figure 8.8b they are convex for 'small' p/c combinations, but as p and c are increased they will 'turn' and become concave like those in figure 8.8a.

The slope of the contract curve is

$$\frac{dp}{dc} = Pf_{cc}(c)[u'(p)]^2/[u(p) - \bar{u}]\, u''(p) < 0 \qquad (8.27a)$$

i.e. the contract curve has a negative slope when $u(p) < \bar{u}$. A graphical illustration is given in figure 8.9. Given the assumed shape of the forest owners' indifference map, we need a few restrictions to ensure the existence of an 'interior' contract curve. Either $(d^2p/dc^2)_{\pi=\bar{\pi}} < (d^2p/dc^2)_{F=\bar{F}}$ must hold as in the figure, or the indifference curves have to be convex over a region around the point of tangency.

The equilibrium contract, as mentioned in section 8.6.3, is a function of the parameters P and k. The dependence is summed up by the following functions, whose partial derivatives have the signs indicated:

$$c = c^s(\underset{+}{P}, \underset{-}{k})$$
$$p = p^s(\underset{+}{P}, \underset{+}{k}) \qquad (8.31)$$

The important difference between the equations in proposition 8.4 and those of (8.31) is that an increased product price increases the roundwood price when the buyer is strong but decreases the roundwood price when the buyer is weak. Moreover, an increased strength on the part of the seller decreases the quantity supplied when the buyer is strong but increases the quantity supplied when the buyer is weak. These differences can, in principle, be used to discriminate between the two cases.

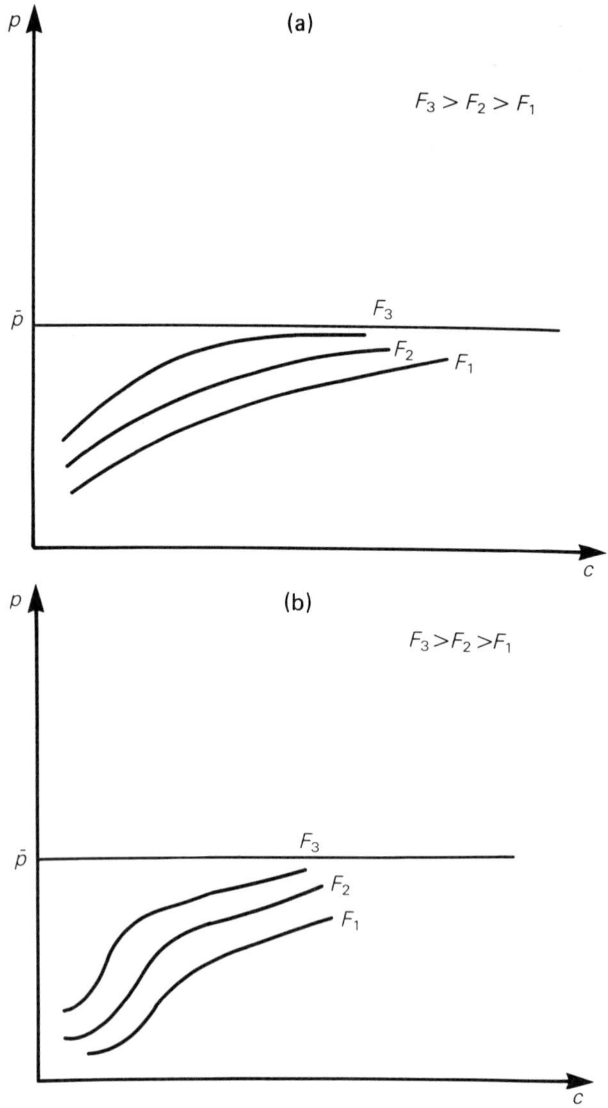

FIGURE 8.8 The indifference curves of the forest owner when $k < 0$

An increased bargaining strength results in higher prices and a lower quantity when the seller is weak because a weak seller who sells below the competitive price wants to sell as little as possible. Increased bargaining strength, therefore, manifests itself in a higher price and a lower quantity supplied.

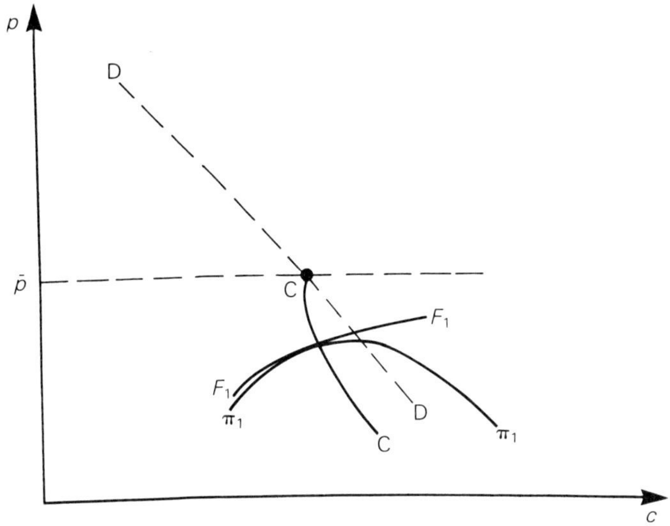

FIGURE 8.9 The contract curve CC when the buyer is strong; $\pi_1\pi_1$ is an iso-profit curve and F_1F_1 an indifference curve

8.6.5 *Bargaining and Price Discrimination*

The case of the strong buyer also gives us an idea of why the forest industry is willing to import roundwood at a price far above the price paid to domestic suppliers. To see this, reformulate the profit function of the forest company as

$$\pi = Pf(c_1 + c_2) - p_1c_1 - p_2c_2$$

where p_1 is the roundwood price paid to domestic suppliers, p_2 is the world market price including cost, insurance, freight (cif), and c_1 and c_2 are the demands for homogeneous roundwood, domestic and imported respectively.

Now assume that the forest company succeeds in discriminating among domestic roundwood suppliers so that $p_1 < \bar{u} \leqslant p_2$. In addition, assume that domestic suppliers face high transport costs so that they are unwilling to sell their roundwood on the world market. Hence the maximization problem

$$\max_{c_1, c_2} L = Pf(c_1 + c_2) - p_1c_1 - p_2c_2 + \lambda\{c_1[u(p_1) - \bar{u}] - k\} \tag{8.32}$$

gives the following first order conditions for an interior maximum:

$$\begin{aligned}
&\text{(a)} \quad \frac{\partial L}{\partial c_1} = Pf_c(c_1+c_2) - p_1 + \lambda[u(p_1) - \bar{u}] = 0 \\
&\text{(b)} \quad \frac{\partial L}{\partial c_2} = Pf_c(c_1+c_2) - p_2 = 0 \\
&\text{(c)} \quad \frac{\partial L}{\partial p_1} = -c_1 + \lambda c_1 u'(p_1) = 0 \\
&\text{(d)} \quad \frac{\partial L}{\partial \lambda} = c_1[u(p_1) - \bar{u}] - k = 0
\end{aligned} \tag{8.33}$$

Condition (8.33b) means that the value of the marginal product of roundwood equals the competitive (import) price (p_2) of roundwood. From (8.33c) it follows that $\lambda = 1/u'(p_1) > 0$, which means that condition (8.33a) implies that the value of the marginal product of roundwood exceeds the price (p_1) which domestic forest owners obtain. In other words, it is profitable for the forest industry to import ('marginal' quantities of) roundwood. This is illustrated in figure 8.10. The curve CC in the figure denotes the contract curve in

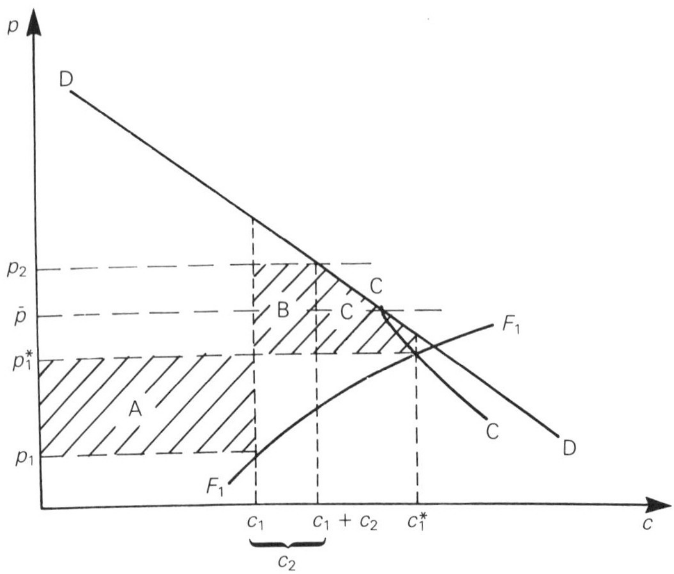

FIGURE 8.10 The use of domestic (c_1) and imported (c_2) roundwood

the absence of imports. If there is no possibility to import roundwood, the forest industry would buy the amount c_1^* at the unit price p_1^* from domestic suppliers. Now let the industry import roundwood at a constant unit price p_2. If this means that the forest industry reduces its purchases of domestic roundwood from c_1^* to, say, c_1, forest owners will be indifferent if the price is reduced from p_1^* to p_1 since combinations (p_1^*, c_1^*) and (p_1, c_1) are on the same indifference curve F_1. However, the forest industry saves $(p_1^* - p_1)c_1$ dollars on the amount c_1. On the other hand, the industry will lose $(p_2 - p_1^*)c_2$ dollars on the amount imported. Note that $Pf_c(c_1) > p_2$ in the figure so it pays to import until $Pf_c(c_1 + c_2) = p_2$. There is also a loss of profit owing to reduced production $[f(c_1 + c_2) < f(c_1^*)]$. Thus the forest industry will reallocate its purchases from the domestic market to the world market as long as the increase in net profit is positive, i.e. as long as the area A in figure 8.10 exceeds the sum of the areas B and C. This model can generate a price pattern which is similar to that actually observed in the Swedish roundwood market during the late 1970s and early 1980s, when Sweden was a net importer of roundwood.

To further illustrate the 'import competition' case, we end this chapter by examining the comparative statics properties of the solution. First of all we note that the price which domestic forest owners receive is determined by the exogenously given world market price of imported roundwood. This can be seen by substituting (8.33b) and (8.33c) into (8.33a). As a consequence, a change in the price of finished products or a change in the 'strength' parameter leaves the roundwood price received by domestic forest owners unchanged. p_1 is affected only through changes in the import price. This is because a change in p_2 affects the value of the marginal product of roundwood, while changes of P and k leave this value constant, as seen in (8.33b). Given p_2, the 'contract curve' – denoted CC in figure 8.10 – will be horizontal. The somewhat surprising result is that if the 'strength' of the forest owners increases, they deliver less roundwood at the same unit price; the price is determined from (8.33b), as noted earlier. To fulfil (8.33d) $\{c_1[u(\bar{p}_1) - \bar{u}] - k = 0\}$ they supply less roundwood.

In short, in the case of imports, the domestic roundwood price and the quantity supplied domestically will depend on the cif price p_2 and the strength parameter in the following manner:

$$\begin{aligned} p_1^{\mathrm{IS}} &= p_1^{\mathrm{IS}}(\underset{+}{p_2}) \\ c_1^{\mathrm{IS}} &= c_1^{\mathrm{IS}}(\underset{+}{p_2}, \underset{-}{k}) \end{aligned} \tag{8.34}$$

where IS denotes the case of the strong buyer when roundwood is imported from the international market. The interesting thing about this extension is that there is no longer any connection between the roundwood price and the price of pulp when there is a net import of roundwood. This hypothesis can be tested by an econometric analysis. The problem is that there is likely to be a strong positive correlation between the price of pulpwood in the international market and the price of pulp.

9

An Econometric Analysis of Aggregate Demand for and Supply of Wood in Sweden

9.1 Introduction

This is a book on theory, and we have spent a lot of time developing a sound microeconomic base for partial equilibrium analysis of the wood market. This chapter is devoted to an econometric analysis of the roundwood trade in Sweden. It can be viewed as an illustration of how some of the 'high powered' theory of chapters 4–8 can be applied to empirical problems. The empirical investigation which we present was originally published in Brännlund, Johansson and Löfgren (1983).

The chapter is structured as follows. In section 9.2 we give a brief outline of the institutional setting in the Swedish roundwood market. In section 9.3 we modify the theory of chapters 6–8 to cover the possibility of different assortments (sawtimber and pulpwood). After a survey of related studies, in section 9.5 we specify the econometric model, and present the estimation results. These are further discussed and qualified in section 9.6. The chapter ends with a summary.

9.2 The Institutional Setting

For the reader's convenience, we repeat some of the institutional details of the Swedish roundwood market. As mentioned in chapter 8, trade occurs in five spatially separated market areas (see section 8.6). Within each market area the price level – the so-called list prices of both sawtimber of different qualities and pulpwood – is set on a yearly basis around August every year in negotiations between one specific area organization representing the sellers and another specific area organization representing the buyers. Interregional trade exists, but is more an exception than a rule. The market where this type of price determination takes place is known as the proper roundwood market, although it is in effect only a national submarket. Buyers within each market area can acquire wood in at least three additional ways. They can supply it from their own forests, they can buy it as standing timber (stumpage), and it can be imported. The possibility of buying timber as standing timber for sale has been a source of supply during the whole period under consideration, while imported wood is a more recent phenomenon.

We model this market structure as being composed of two interrelated markets, a sawtimber market and a pulpwood market, with one demand curve and one supply curve in each. We are mainly interested in the properties of the supply curves, but to identify the supply curves we need some kind of model of how the demand side works and/or how prices are formed. Roughly speaking, we choose to model the sawtimber market as a market where demand and supply determine the price level, while the market for pulpwood is viewed as a market where the price level is exogenously determined, i.e. determined without any interaction between demand and supply. Hence, the actual supply of pulpwood in a particular period can be identified without knowledge of the demand conditions in that particular period. We will later elaborate on the particular model in the theoretical section.

The results of this exercise are satisfactory. All coefficients of the supply curves have the expected signs and most of them are statistically significant. The same holds true for the demand curve in the sawtimber market. This is somewhat surprising since conventional wisdom in the field sometimes describes the wood trade in Sweden as something quite irrational, or compares it with horsetrading (whether the latter is irrational or not is a question which we are no longer competent to answer).

9.3 The Demand for and Supply of Wood and the Partial Multimarket Equilibrium

Wood is used as an input in the production of sawnwood or in the processing of pulp and paper. We assume that sawnwood is sold in the international market at a constant price p_t^{sw}, and that the final product is produced by using the twice continuously differentiable production function

$$Q_t^{\mathrm{s}} = F(l_t^{\mathrm{d}}, c_t^{\mathrm{d}})$$

where l_t^{d} is the input of labour, c_t^{d} is the input of wood, Q_t^{s} is output of sawnwood and t is the period ($t = 1, \ldots, T$).

We can then define the profit function in period t as

$$\pi(p_t^{\mathrm{sw}}, p_t, w_t) = \max_{c_t^{\mathrm{d}}, l_t^{\mathrm{d}}} p_t^{\mathrm{sw}} F(l_t^{\mathrm{d}}, c_t^{\mathrm{d}}) - p_t c_t^{\mathrm{d}} - w_t l_t^{\mathrm{d}} \tag{9.1}$$

where w_t is the wage rate, p_t is the price of roundwood and p_t^{sw} is the price of sawnwood.

By applying Hotelling's lemma, we obtain the demand for roundwood in the tth period by differentiating the profit function with respect to p_t. This yields

$$\pi_{p_t} = c_t^{\mathrm{d}}(\underset{(+)}{p_t^{\mathrm{sw}}}, \underset{-}{p_t}, \underset{(-)}{w_t}) = c_t^{\mathrm{d}}\left(\frac{p_t^{\mathrm{sw}}}{w_t}, \frac{p_t}{w_t}\right) \tag{9.2}$$

The signs under the arguments denote the signs of the partial derivatives. The second equality in (9.2) follows from homogeneity of degree zero of the demand function. The parentheses under two of the arguments indicate that the signs are correct if labour and roundwood are technical complements:

$$F_{cl} = F_{lc} > 0$$

This analysis of demand – like the analysis in the preceding chapters – is conducted in a static setting, but it is easy to imagine how the demand for inputs in an intertemporal setting would depend on future prices. To preserve simplicity, we assume that a firm's demand for wood in a particular period t ($t = 1, \ldots, T$) depends only on the prices of that period.

Using the results from chapter 8, the aggregate demand for wood is obtained by summing the individual firms' demand for wood:

$$C_t^{\mathrm{d}} = \sum_{j=1}^{J} c_{tj}^{\mathrm{d}}(p_t^{\mathrm{sw}}, p_t, w_t) \qquad \text{all } t \tag{9.3}$$

The aggregate supply of roundwood is obtained in a similar manner by aggregating individual supply functions, i.e.

$$C_t^{\mathrm{s}} = \sum_{h=1}^{H} c_{th}(p) \qquad \text{all } t \tag{9.4}$$

where $p = (p_1, \ldots, p_T)$ is a vector of roundwood prices [see (8.3)].

A partial multimarket equilibrium is defined as

$$C_t^{\mathrm{d}}(p_t^{\mathrm{sw}}, p_t^*, w_t) = C_t^{\mathrm{s}}(p^*) \qquad \text{all } t \tag{9.5}$$

i.e. as the equality between supply and demand in each period; $p^* = (p_1^*, \ldots, p_t^*, \ldots, p_T^*)$ is the equilibrium roundwood price vector. Note that p_t^{sw} and w_t are assumed to be exogenously determined.

There are two problems with the use of the formulation in (9.5) for an empirical specification of the model. One is that two different categories of products are traded in the Swedish roundwood market – pulpwood and sawtimber. A second problem is that while it is not unreasonable to model the Swedish sawtimber market as a perfect market with many sellers and many buyers, the pulpwood market in Sweden is characterized by many sellers but few buyers. This indicates that the pulpwood market is a monopsonistic market.

To move from the LNF model in chapter 6, with a single output (wood), to a model with two product categories (pulpwood and sawtimber), we rewrite the present value of the cut as

$$\Pi(c^s, C^s, l) = p^p c^s + p^s C^s - wl \tag{9.6}$$

where $p^p = (p_1^p, \ldots, p_T^p)$ is a vector of discounted pulpwood prices, p^s of discounted sawtimber prices, w of discounted wages, c^s of pulpwood supplies, C^s of sawtimber supplies and l of labour demands. In addition we introduce the set

$$G = \{c, c^s, C^s\colon g_1(c(1), c_1^s, C_1^s) = 0, \ldots, g_T(c(T), c_T^s, C_T^s) = 0\} \tag{9.7}$$

The implicit functions $g_t(c(t), c_t^s, C_t^s) = 0$ are the feasible possibilities to transform a given composition of the cut $c(t)$ into two final products, pulpwood c_t^s and sawtimber C_t^s. The present value function can now be written [see (6.28)]

$$\Pi(p^p, p^s, w, \bar{x}_0) = \max_{c^s, C^s, c^a, l} \{p^p c^s + p^s C^s - wl\colon c^s, C^s, c^a, l \in F \cap G \cap C\} \tag{9.8}$$

where

$$F\{c^a, l\colon F(c^a, l) = 0\}$$
$$C = \{c^a\colon c^a \text{ satisfies } (6.15)\}$$
$$c^a = [c(1), \ldots, c(T)]$$
$$\bar{x}_0 = (\bar{x}_{00}, \ldots, \bar{x}_{0n})$$

From Hotelling's lemma (proposition 6.9) we know that the supply curves of pulpwood and sawtimber can be obtained by differentiating the present value function with respect to p^p and p^s, respectively:

$$\frac{\partial \Pi}{\partial p^p} = c^s(p^p, p^s, w, \bar{x}_0)$$

$$\frac{\partial \Pi}{\partial p^s} = C^s(p^p, p^s, w, \bar{x}_0) \tag{9.9}$$

This means that the equilibrium condition (9.5) for the *sawtimber* market is modified to read

$$C_t^d(p_t^{sw}, p_t^s, w_t) = C_t^s(p^s, p^p, w, \bar{x}_0) \qquad \text{all } t \tag{9.10}$$

In line with the actual pattern of price formation in the Swedish *pulpwood* market, it is assumed that pulpwood prices are set by a monopsonistic paper and pulp industry, which bases its optimization calculation on an expected supply curve $c_t^e(p_t^p; p_{t-1}^s, w)$. The industry maximizes profit:

$$\max_{p_t^p} \Pi^p = P_t^p f[c_t^e(p_t^p; p_{t-1}^s, w)] - p_t^p c_t^e(p_t^p; p_{t-1}^s, w) \qquad \text{all } t \tag{9.11}$$

which yields the pulpwood price in period t as a function of the world market price of pulp P_t^p and the lagged sawtimber price. (For a more rigorous approach based on explicit stochastic elements the reader is referred to Löfgren and Ranneby, 1984.)

The pulpwood price is given to the market from outside, i.e. actual supply is determined by the pulpwood supply functions

$$c_t^s = c_t^s[p^s, p^p, w, \bar{x}_0] \qquad \text{all } t \tag{9.12}$$

where all price vectors are exogenously determined. In section 9.4 we estimate a somewhat simplified version of the system (9.10)–(9.12).

9.4 Previous Studies

Many of the previous econometric studies on timber supply have been concerned with North America. Robinson (1974) estimated the supply elasticity from Douglas-Fir with an econometric model of softwood lumber and stumpage markets. He found a fairly low supply elasticity. A similar investigation was made by Adams (1977) who, in an analysis of harvesting behaviour of private forest owners in the US, also found a low timber supply elasticity ranging between 0.2 and 0.5.

Binkley (1979) estimated the probability of harvest in a paper covering harvesting behaviour of private forest owners in New Hampshire. The results indicated that the probability of timber harvest is strongly affected by stumpage price. It would, however, be too hasty to say that the volume of the harvest is strongly affected by stumpage prices, since a more frequent harvest is not equivalent to a larger harvest.

In Sweden, the supply behaviour of private non-industrial forest owners has been scrutinized in several so-called 'soft studies', most of them executed by Lönnstedt (see e.g. Lönnstedt, 1977). An early econometric study was conducted by Ruist and Svennilson (1948).

The timber supply behaviour of the private non-industrial forest owners in Finland has recently been analysed in a probit model by Kuuluvainen, Loikkanen and Salo (1983). Sales probability was regressed on various qualitative variables. They found, as expected, that the sales probability of farmers was much higher than for non-farmers, and that an 'education in forestry variable' significantly affected only the cutting decision for non-farmers. However, Seppälä has published two papers (1974a, b) on the supply behaviour of private forest owners in Finland, and in one of them demonstrated that supply is positively correlated with wood prices.

9.5 The Empirical Model and the Empirical Results

From the earlier theoretical analysis we get a very good idea of how to specify the empirical model. We divide the analysis of roundwood supply into an analysis of two distinct but interrelated markets – the market for sawtimber and the market for pulpwood. The demand side of the timber market is explicitly specified by a demand function with the price of sawnwood, the price of timber, and the wage rate in the sawmill industry as arguments. The supply side of the market consists of a supply function, which in a minimum version of the model contains the price of timber, the price of pulpwood, and cutting costs as arguments [corresponding to w in (9.10)]. The reasoning behind the use of pulpwood prices as a separate argument is that some of the smallwood could be used either as sawtimber or as pulpwood. The higher the price of pulpwood, the more smallwood is transferred from the timber market to the pulpwood market. There is, in addition, an effect which stems from the fact that a higher price for pulpwood makes thinning more profitable than before (unprofitable thinnings become profitable), and this persuades some of the forest owners to carry out a thinning instead of a final felling, which in turn decreases the supply of sawtimber and increases the supply of pulpwood.

The pulpwood market is treated in a somewhat different manner. As was indicated earlier, there are theoretical and institutional reasons why the price of pulpwood should be treated as an exogenous variable, i.e. not determined by the simultaneous interplay between supply and demand, in an analysis of pulpwood supply. Since the industry is the major party in the price negotiations between buyers and sellers, the price can be viewed as 'determined' from the demand side of the market. The price of pulpwood is then a reflection of the expected needs of wood in the pulp and paper industry, and the forest owners react to the given price by adjusting supply accordingly. We estimate the supply of pulpwood as a function of the price of pulpwood, the cutting costs, and the price of sawtimber. The latter argument is used to cover the possibilities of smallwood as either sawtimber or pulpwood. Hence, an increased sawtimber price should decrease the supply of pulpwood.

We have estimated linear and log-linear versions of this standard model and most estimates provide a reasonably good fit (see Brännlund, Johansson and Löfgren, 1983). One problem, however, is that the demand curve for sawtimber – according to casual empiricism – seems to have too high a price elasticity. An increase in the sawtimber price by 10 per cent would 'choke off' 17 per cent of the demand for sawtimber, while an increase in the unit value of sawnwood by 10 per cent would increase demand by 15 per cent.

We have chosen to present a minor modification of the standard model. In the equation for the demand for sawtimber, the price for sawnwood and the wage rate are merged to form a variable which we call the ability to pay for sawtimber. We have also introduced lagged production of sawnwood as a separate argument. One way to look upon the latter argument is to say that the demand for wood in period t depends not only on prices of the final product in the same period, but also on the expectations of future prices which were created a period ago. Another related way to view it is that adjustment lags make the demand for wood in a particular period an increasing function of the level of sawnwood production in the preceding period.

The partial equilibrium model of the sawtimber market can hence be written

$$
\begin{aligned}
C_t^{\mathrm{D}} &= \alpha_0 + \underset{(-)}{\alpha_1 p_t^{\mathrm{s}}} + \underset{(+)}{\alpha_2 (p^{\mathrm{sw}} - w^{\mathrm{s}})_t} + \underset{(+)}{\alpha_2 Q_{t-1}} \\
C_t^{\mathrm{s}} &= \beta_0 + \underset{(+)}{\beta_1 p_t^{\mathrm{s}}} + \underset{(-)}{\beta_2 p_t^{\mathrm{p}}} + \underset{(-)}{\beta_3 Z_t} \qquad (9.13\mathrm{a}) \\
C_t^{\mathrm{D}} &= C_t^{\mathrm{s}}
\end{aligned}
$$

where the variables have the following meanings

C_t^{D}	demand for sawtimber
C_t^{s}	supply of sawtimber
p_t^{sw}	the world market prices of sawn products
p_t^{s}	price of sawtimber
p_t^{p}	price of pulpwood
w_t^{s}	wage rate in the sawmill industry
Z_t	cutting cost [corresponds to the w vector in (9.10)]
$(p^{\mathrm{sw}}-w^{\mathrm{s}})_t$	ability to pay for sawtimber
Q_{t-1}	lagged sawnwood production

All prices and costs are measured in real terms. We have used the consumer price index as a deflator, but the estimates are robust with respect to the choice of a deflator, owing to the strong correlation between practically feasible deflators. The signs under the arguments denote the expected signs of the coefficients.

In the supply equation for the pulpwood market, we have tried to model the possibility that today's price as well as the expected future prices affect the supply of pulpwood c_t^{s}, i.e.

$$c_t^{\mathrm{s}} = \gamma_0 + \underset{(-)}{\gamma_1 p_t^{\mathrm{s}}} + \underset{(+)}{\gamma_2 p_t^{\mathrm{p}}} + \underset{(-)}{\gamma_3 Z_t} + \underset{(-)}{\gamma_4 p^{\mathrm{e}}} \tag{9.14}$$

where p^{e} represents the expected future price level. The expectations of future prices are assumed to be based on a distributed lag of present and previous prices:

$$p^{\mathrm{e}} = \sum_{i=0}^{\infty} \beta_{t-i} p_{t-i}^{\mathrm{p}} \tag{9.15}$$

Substitution of (9.15) into (9.14) yields, after appropriate transformations, the following possible specification of the supply equation:[1]

$$c_t^{\mathrm{s}} = \gamma_0' + \underset{(-)}{\gamma_1' p_t^{\mathrm{s}}} + \underset{(+)}{\gamma_2' p_t^{\mathrm{p}}} + \underset{(-)}{\gamma_3' Z_t} + \underset{(-)}{\gamma_5 p_{t-1}^{\mathrm{p}}} + \underset{(+)}{\gamma_6 c_{t-1}^{\mathrm{s}}} \tag{9.13b}$$

The effect of an increase in today's price of pulpwood on the supply of pulpwood should be reduced by the introduction of explicit expectations of future prices. This follows in a deterministic manner from the homogeneity of degree zero in the prices of the supply function, and the distributed lag character of the expectations mechanism ($\gamma_4 < 0$).

[1] p^{S} and Z should also appropriately enter as lagged variables, but the respective coefficients are insignificant, and the variables have therefore been taken out.

9.5.1 *Data*

The data used are annual time series covering the period 1953 to 1981. The data sources are described in Brännlund, Johansson and Löfgren (1983). The price data on sawtimber prices and pulpwood prices cover the whole country, and are the so-called list prices. The potential weakness of list prices is discussed in section 9.6.

We use aggregate data mainly because it is not possible to obtain data for each separate market area. We report, however, some preliminary estimates using data from price area 1 – the northern part of Sweden.

Figure 9.1 shows the combinations of cutting behaviour and prices that our model is designed to explain. The cutting volumes refer to the whole country, and the prices are yearly averages for the entire country. The points in figure 9.1 should be viewed as the loci of intersections between supply and demand curves. The positions of these curves in the price–quantity plane are determined by the pulpwood price, the cutting cost, the price of sawnwood and the production costs in the sawmill industry.

9.5.2 *The Results of the Estimations*

Linear and log-linear variations of the simultaneous model consisting of (9.13a) and (9.13b) have been estimated by two stage least squares (2SLS), ordinary least squares (OLS) and three stage least squares (3SLS) (see e.g. Pindyck and Rubinfeld, 1976). The residuals have been treated as being generated by an autoregressive process of first order (AR-1) process, according to the Cochrane-Orcutt iterative technique.

The model has a causal ordering of endogenous variable coefficients necessary for estimating the pulpwood supply equation by ordinary least squares. However, since sawtimber and pulpwood harvest decisions are probably made contemporaneously, the sufficient condition for using OLS, that the variance covariance matrix is diagonal, may fail to be satisfied. If this is the case the appropriate estimation scheme is to apply 3SLS to the entire system of equations including the pulpwood equation.

We have, however, chosen to present only the OLS estimate of the pulpwood equation. The reason is that the 3SLS estimates show only minor differences from the OLS estimates, while the Durbin-Watson (DW) statistic indicates first order autocorrelation, for which we cannot correct under 3SLS.

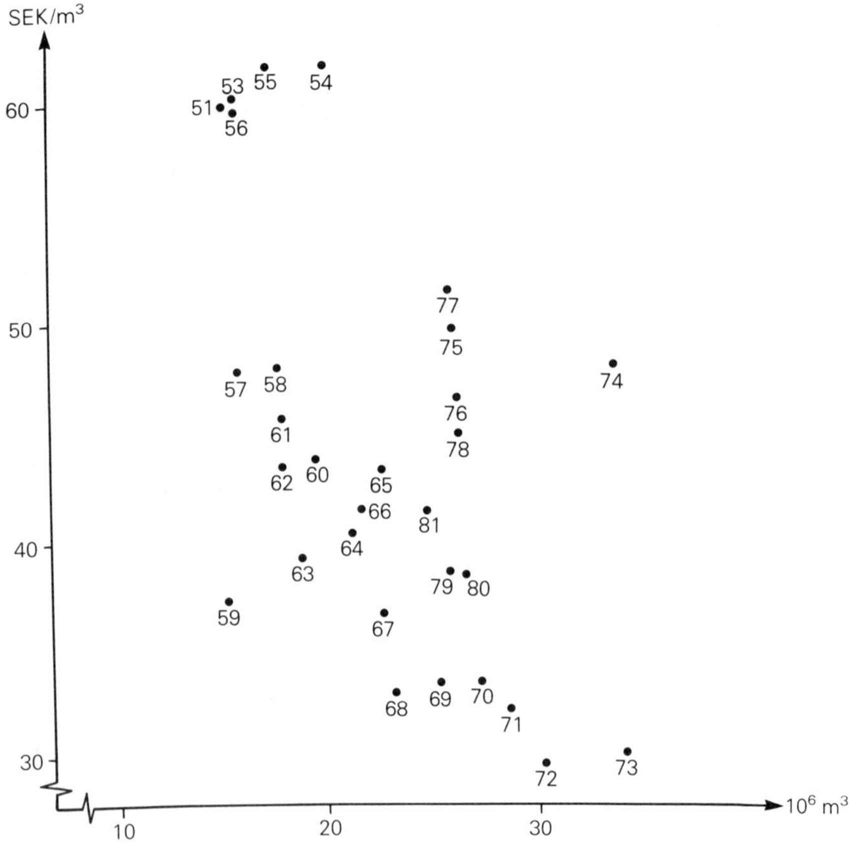

FIGURE 9.1 Sawtimber prices and production (stem volume over bark) in Sweden, 1951–81

The 2SLS estimates of the demand and supply curve of sawtimber are presented in table 9.1. The model is log-linear, so the coefficients can be interpreted as elasticities. As can be seen from the table, the model has a good fit. All coefficients in the supply curve are statistically significant and have the expected signs. The estimates tell us that the supply elasticity of an increase in the sawtimber price is 0.609, i.e. an exogenous shift of the demand curve which increases the sawtimber price by 10 per cent will increase the supply of sawtimber by approximately 6 per cent. The coefficient for the pulpwood price tells us that a 10 per cent increase in the pulpwood price will decrease the supply of sawtimber by approximately 6 per cent. Taken together these results mean that it is difficult to increase

TABLE 9.1 The supply and demand functions for sawtimber (log-linear version): modified model

	C_t^{S}	C_t^{D}
α_0	5.283**	−4.957**
p_t^{s}	0.609*	−0.989**
p_t^{p}	−0.681**	
Z_t	−0.848**	
$(p^{\text{sw}} - w^{\text{s}})_t$		0.852*
Q_{t-1}		0.848**
R^2	0.998	0.998
DW	1.31	2.03

* Significant at the 5 per cent probability level.
** Significant at the 1 per cent probability level.

TABLE 9.2 The reduced form estimates of the cut and the price of sawtimber

	$\bar{C}_t$	$\bar{p}_t^{\text{s}}$
α_0	−1.19	−3.712**
p_t^{p}	−0.417**	0.430**
Z_t	−0.308	0.279
$(p^{\text{sw}} - w^{\text{s}})_t$	0.302*	0.554**
Q_{t-1}	0.553	0.294*
R^2	0.85	0.88
DW	1.71	2.32

* Significant at the 5 per cent probability level.
** Significant at the 1 per cent probability level.

sawtimber supply by price stimuli during periods in which the business cycles of the pulp and paper industry and the sawmill industry run in 'phase' with each other. Finally, an increase in cutting costs has the expected negative influence on sawtimber supply.

The demand elasticities have a reasonable size, the coefficient for the ability to pay has the expected sign, and the lagged production variable (Q_{t-1}) has both a reasonable sign and a reasonable size. All coefficients in the demand equation are strongly significant. The reduced form estimates of both the cut and the price of sawtimber are given in Table 9.2.

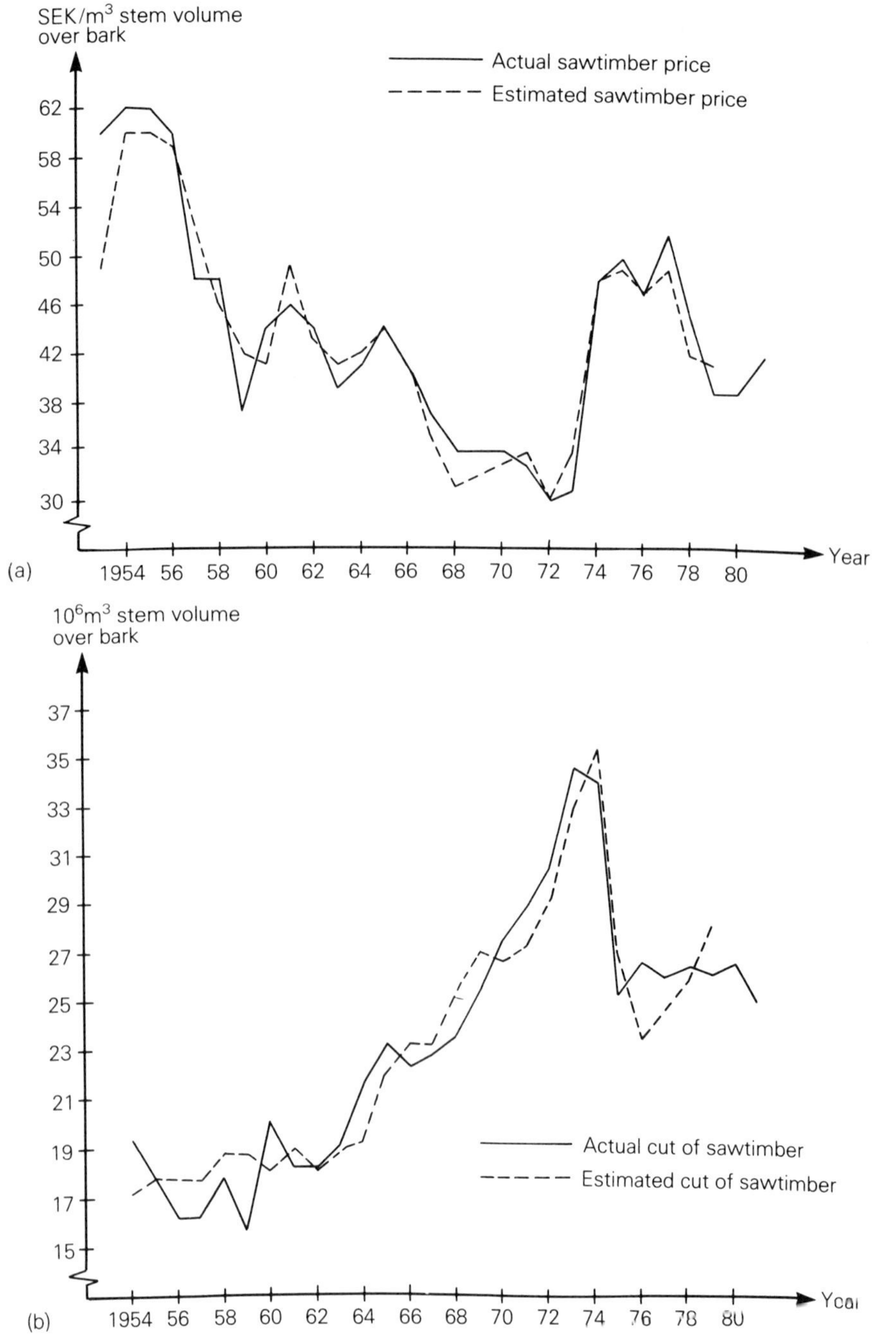

FIGURE 9.2 Actual and fitted (a) prices and (b) quantities

TABLE 9.3 The supply function of pulpwood

	c_t^s
γ_0'	2.61**
p_t^p	0.744**
p_{t-1}^p	−0.722**
p_t^s	−0.283
Z_t	−0.089
c_{t-1}^s	0.591**
R^2	0.84
Durbin h	0.06

* Significant at the 5 per cent probability level.
** Significant at the 1 per cent probability level.

Figure 9.2 shows the predictive power of the model. The predictions of sawtimber price and cut for the period 1953–80 have been obtained by making use of the corresponding reduced form equations. It could be argued that it would be more appropriate to use a reduced form obtained by solving the structural equations. However, when this is undertaken, it turns out that nothing essential is changed; the fit remains satisfactory.

The model is able to describe the fluctuations in prices and quantieis in a satisfactory manner. It is worth noting that the model picks up the steep decline of the cut from 1973–5. However, the model overestimates the supply during the end of the period. This is interesting, because at the end of this period there were some indications of rationing in the Swedish roundwood market, i.e. the amount of wood buyers were willing to buy at the ruling price was larger than the amount supplied. If we are to believe the model, these tendencies could have been pronounced in 1979, as the model in this year overestimates supply by 2 million m^3.

The estimates of the supply function for pulpwood are presented in Table 9.3. All coefficients have the expected signs. The R^2 value is reasonable and this is also true for the Durbin h.[2] However, both the coefficient for cutting cost and the coefficient for the sawtimber price are not statistically significant. The direct supply elasticity is slightly lower than the supply elasticity in a standard model – 0.74

[2] Used as an indicator of autocorrelation, instead of the DW statistic, when the equation to be estimated contains the lagged dependent variable.

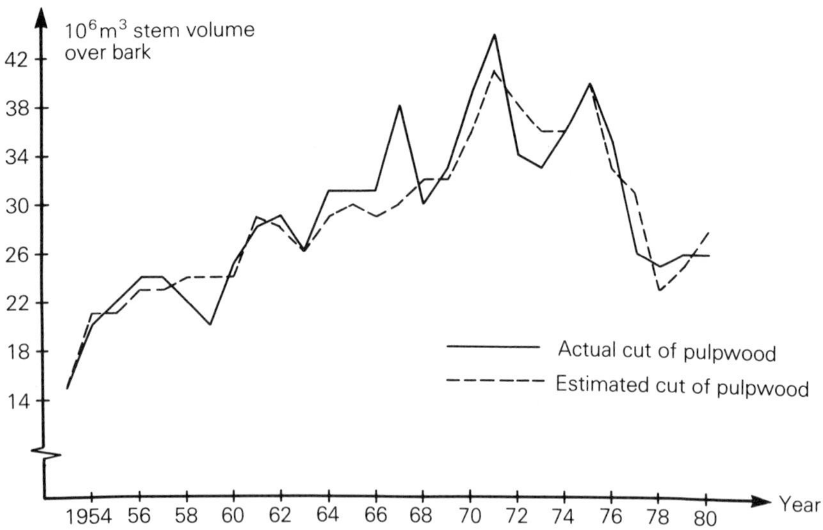

FIGURE 9.3 Actual and fitted supply of pulpwood

compared with 0.85 – which was to be expected in a model containing a variable which picks up price expectations. It is worth noting the small influence of the sawtimber price on pulpwood supply. A possible explanation is the following. The sawtimber price is likely to have two effects on pulpwood supply. One is due to the fact that an increased sawtimber price increases the relative profitability of final cuttings in relation to thinnings, and this has a negative effect on pulpwood supply. The second effect is a 'joint production effect'; every increase in final cuttings will produce an increased volume of both sawtimber and pulpwood, and this effect counteracts the negative substitution effect. Returning to the estimates of the sawtimber supply curve, we find that the pulpwood price has a considerable negative influence on the supply of sawtimber. Interpreted in the decomposition, it should be clear that an increased pulpwood price has a negative substitution effect, but also a negative joint production effect since thinnings result in mainly pulpwood. This explains the obvious asymmetry in the estimated cross-price effects.[3]

If we allow the model to predict pulpwood supply during the period under consideration, we obtain the picture in figure 9.3. With the exceptions of the decline during the period 1957–9 and the peak in 1967, the model shows most of the variations in pulpwood supply.

[3] From proposition 6.10 we know that this asymmetry indicates a discontinuity of the present value function.

9.6 Discussion

Does the level of aggregation disguise large qualitative behavioural differences between different kinds of owners of forest land? About one half of the forest land in Sweden is owned by non-industrial private owners, one quarter is owned by the state, and the remaining quarter is owned by the forest industry. It has been argued that the cutting behaviour of the industry is negatively correlated with the harvesting behaviour of private forest owners, in the sense that a large harvest by private forest owners is combined with a small harvest by the industry and vice versa.

This conjecture cannot, however, be supported by economic theory. The economic theory of chapter 7 tells us that the industry's opportunity cost of wood in a perfect market equals the market price, and that the opportunity cost in an imperfect market is positively correlated with the market price of wood. The reason for this is that the industry has a choice between buying the timber in the market or harvesting it in its own forest. Hence, it is obvious that a high market price makes it particularly favourable to cut wood from one's own forest. Technically speaking, this means that the positively sloped supply function is also relevant for timber supplied from forest land owned by industry. This is certainly true for wood supplied from state lands as a profit goal is required of the forest service. (Corollary 6.2 is, hence, valid.)

These conclusions are supported by figure 9.4 where the total harvest has been disaggregated into two categories – the total cut by industry and state and the total cut by private forest owners. The cutting patterns show the expected qualitative similarity. It would, nevertheless, be very interesting to measure the quantitative differences between different kinds of owners, which are for example crucial to certain issues of forest policy, such as the response to tax reforms. Our data source does not allow this urgent venture because it is impossible to separate sawtimber from pulpwood at a lower level of aggregation. Some interesting work in this direction using cross-section data is presently under way.

Another potential weakness of our results is the price data, since only part of the roundwood trade is conducted at these prices. As the level of prices is 'uninteresting' in itself (statistics is a question of variation), we have correlated our list prices with a series of prices of standing timber for sale. (This series, which is obtained from the *Statistical Yearbook of Forestry*, is also incomplete as it covers only part of the stumpage market.) The correlation coefficients are 0.81

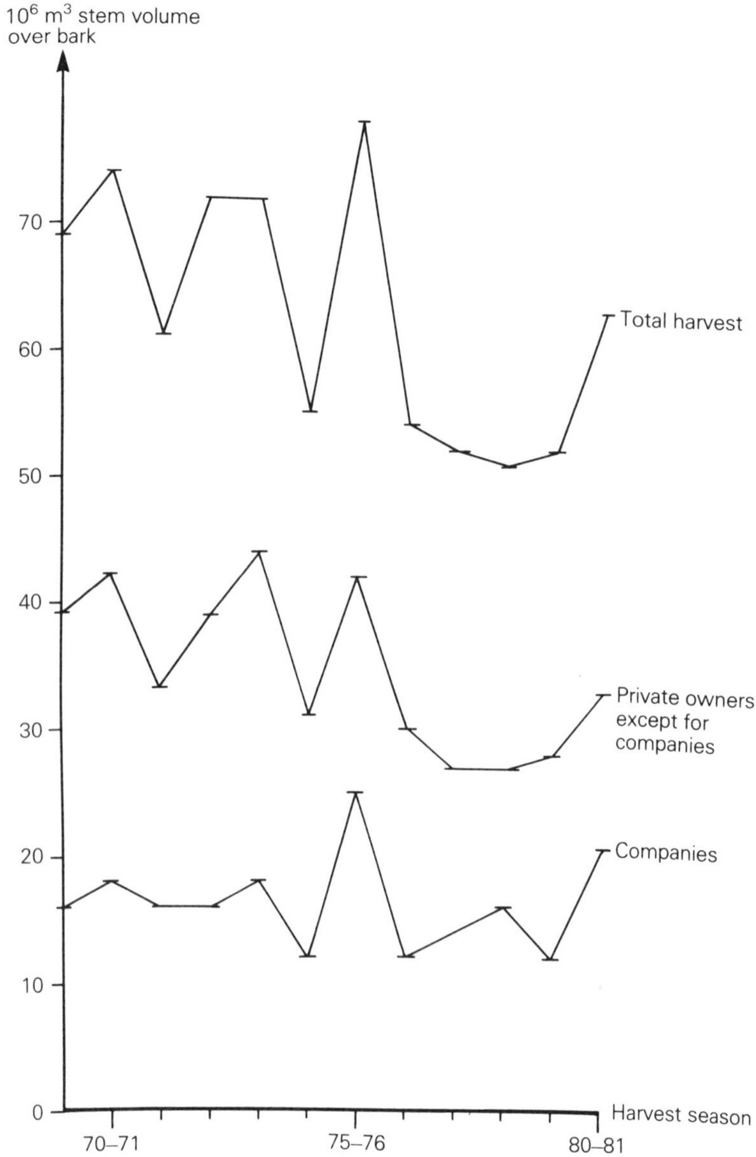

FIGURE 9.4 Disaggregation of the total harvest in Sweden, 1970-81

for pulpwood prices and 0.89 for sawtimber prices, which, to some extent, is comforting evidence about the validity of our data set. Another comforting fact is that the list prices are used as accounting prices when industry firms swap timber and pulpwood.

One could argue on theoretical grounds that some kind of proxy for the rate of return (the discount rate) should enter the supply function as an argument; remember that the prices in chapters 4–6 are discounted prices. We have tried the net (after tax) real interest rate on bonds and/or the real yield from equities, but the variables were – in most runs – not statistically significant, and generated coefficients with incorrect signs. We prefer to view these results as an indicator that bonds and equities are not the relevant assets for a forest owner. Other interpretations are of course quite possible.

The theoretical analysis in chapter 6 suggests the inclusion of a proxy for the composition of the initial endowment. We have tried different proxies without any appreciable success. Certain readers might also consider that some kind of tax variable is missing from the supply equation. We would of course have liked to include one, but there are a lot of technical problems involved. One is that different groups face different tax rates and that, although Swedish tax rates are high, they are constantly so!

9.7 Concluding Comments

One of the problems in describing the cutting behaviour of Swedish forest owners is that there are many different kinds of owners facing different kinds of economic conditions. This line of argument is, however, incomplete since it does not preclude the possibility that agents in different economic conditions react in the same manner to the same kind of economic stimulus. If we believe our results, this is to some extent the case. Nevertheless, it would be interesting to measure quantitatively the differences in price elasticities between different ownership categories, such as non-industrial private forest owners and industrial forest owners.

The estimates of timber supply presented show that Swedish forest owners seem to react to price stimuli, and also that timber supply can be modelled by using a simple model of a perfect market. The latter is rather surprising since there has been a tendency for rationing and price discrimination to become prevalent at least towards the end of the investigated period. However, a closer look at figures 9.2 and 9.3 also reveal that the fit of the model deteriorates at the end of the period. Both sawtimber and pulpwood supply are overestimated by the model in the years 1979–80.

Another feature of the model worth mentioning is the inelasticity of sawtimber supply under simultaneous increases in both sawtimber prices and pulpwood prices. According to the model, a simultaneous increase[4] in these prices by, say, 10 per cent would slightly decrease sawtimber supply. One reason behind this phenomenon could be related to the substitution effect discussed earlier. An increase in the pulpwood price will mean that formerly unprofitable thinnings become profitable, and this induces a transformation of felling capacity from final cuttings to thinnings. Another possible reason is the homogeneity of degree zero of the supply function (proposition 6.6). If a 10 per cent general increase in the price of wood induces the forest owner to believe that all future prices will increase by 10 per cent this leaves the optimal cutting programme unaffected. A third reason could be risk aversion in connection with the uncertainty of future prices, which can be shown to make the slope of the supply curve ambiguous. See the analysis in chapter 12. A fourth reason, finally, could be the presence of constraints which regulate the permissible cut. It can be shown that such regulatory constraints make the supply function less elastic (see Cuddington, Johansson and Löfgren, 1984, appendix A and Johansson and Löfgren, 1983b).

To discriminate between these possible explanations or to select the most important one is far from easy. Nevertheless, an affirmative answer to what is the most important reason for the inelastic supply would be interesting.

The model is estimated on data for the whole country. This is probably a clear advantage for the fit of the model, as irrational and/or extraordinary reasons to cut tend to cancel out. It would, however, be interesting to estimate the model on data from some of the regionally separated price areas. Some preliminary estimates on data from price area 1 – the northern part of Sweden – indicates that the precision of the estimates deteriorates somewhat, while the general trends prevail.

[4] Caused by some exogenous shock, e.g. an increased ability to pay for sawtimber, which shifts the demand function outwards generating a higher sawtimber price. The reader should note that it is not meaningful to talk about the impact of an increased price of sawtimber on the supply of sawtimber, both are simultaneously determined.

10

Welfare Theory and the Optimal Pricing of Natural Resources

10.1 Introduction

The purpose of this chapter is to show how traditional welfare theory can be applied to find a Pareto optimal pricing plan for natural resources. We touched upon questions of welfare theory in previous chapters. In section 3.3 the problem of the commons was analysed fairly rigorously, and in the appendix at the end of chapter 3 the management of a renewable natural resource, which has value to society both when it is consumed and when it is present as a stock, was discussed. In section 4.5 we derived the socially optimal rotation period.

In this chapter the welfare problem, and in particular the optimal pricing problem, is given a more thorough analysis. Production conditions and biological growth functions are less elaborate than previous chapters, but the social optimization problem is stated in a more general fashion. We use the model of Baumol and Oates (1975), and our propositions, with a few important exceptions, can be found in their excellent book on environmental policy (see also Clawson and Knetsch, 1966; Herfindahl and Kneese, 1974; and Mäler, 1974). The analysis of welfare theoretical issues continues in chapter 11, where we derive cost–benefit rules for natural resources.

10.2 Pareto Optimal States

Consider a simple economy where commodity q is produced with labour. To avoid notational difficulties, we assume that there is only one plant with the strictly concave and twice continuously differentiable production function

$$q = f(l) \tag{10.1}$$

where

$$l = \sum_{i=1}^{N} l_i$$

is the total amount of labour supplied by the N individuals in the economy.

There are N individuals in the economy and each individual possesses a strictly quasi concave and twice continuously differentiable utility function U_i with consumption q_i and leisure r_i as arguments, i.e.

$$\left.\begin{aligned} U_i &= U_i(q_i, r_i) \\ r_i &= L - l_i \end{aligned}\right\} \quad \text{all } i \tag{10.2}$$

where q_i is the quantity of commodity q consumed by individual i, l_i is the labour supplied by individual i, r_i is the amount of leisure consumed by individual i and L is the total amount of available time.

We will use the terms *Pareto optimum* and *social optimum* synonymously. A Pareto optimum is defined in the following manner:

Definition 10.1
A state E^0 is called a Pareto optimum if it is feasible, and if there exists no other state E^1 such that

$$U_i(q_i^1, r_i^1) \geqslant U_i(q_i^0, r_i^0) \qquad \text{all } i$$

where strict inequality holds for at least one individual.

In other words, E^0 is Pareto optimal if it is possible to reach it given the resource constraints and if, given E^0, the utility of one consumer cannot be increased without decreasing the utility of at least one other consumer. The set of Pareto optimal states may contain more than one element, i.e. there is a multiplicity of optimal states. Figure 10.1 illustrates the set of feasible states in an economy with two consumers; their levels of utility are denoted U_1 and U_2 respectively. All feasible states on the boundary of the feasible region and to the right of point A are clearly Pareto optimal.

We want to find a Pareto optimal state in the model economy defined by (10.1) and (10.2). One method is to maximize the utility of one individual, subject to all other individuals being kept at a given feasible utility level and given the production function and resource constraints. This, however, does not always work, as illustrated by point B in figure 10.1. Maximizing the utility of individual 1, given that individual 2 is kept on the utility level $\bar{U}_2$, yields point B, which is clearly not Pareto optimal since the utility of both individuals can be improved by a move to point A.

This method works better if the utility levels of other individuals are allowed to exceed given utility levels. For all levels $\bar{U}_2 \leqslant \bar{U}_2^A$ in figure 10.1 the maximization problem is solved by the allocation A, and for utility levels $\bar{U}_2 > \bar{U}_2^A$ the solution is somewhere on the curve AC, where all points are Pareto optimal. In fact, this method always gives a unique solution as long as the boundary of the feasible region has no horizontal or vertical segments. Another way to avoid odd cases is to define a Pareto optimal state as a state from which it is impossible to increase the utility of all individuals. Under this new definition the method works even if the boundary has horizontal or vertical segments.

10.3 Optimal Pricing of Non-renewable and Renewable Resources

We now apply this welfare theoretic technique to the efficiency problem for a model economy which contains natural resources. As

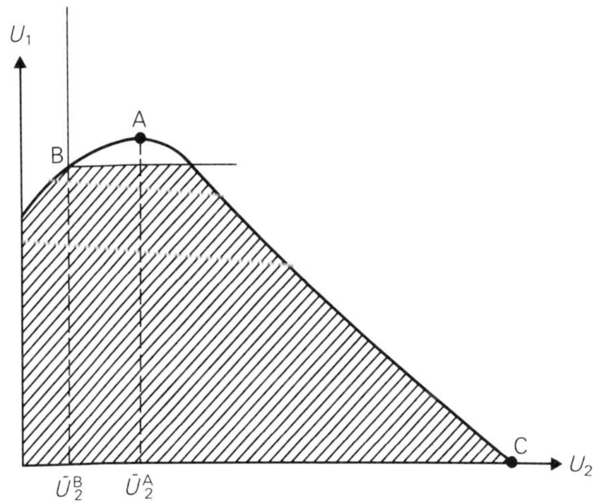

FIGURE 10.1 The set of feasible states in the utility space

mentioned in section 10.1 the particular model is due to Baumol and Oates (1975). The basic model consists of the following variables:

x_{jt}	quantity of the (renewable) resource X consumed by individual j in period t, where $j = 1, 2, \ldots, N$ and $t = 1, 2, \ldots, T$
x_t	unconsumed quantity of the resource X at time t
x_0	initial stock of the (renewable) resource X
k	rate of growth of the resource X
q_{jt}	quantity of the second good (Q) consumed by individual j during period t
q_t	total output of commodity Q in period t
$\bar{q}_t$	unconsumed quantity of Q remaining at the end of t
r_{jt}	leisure demanded by individual j during t
L	total labour time available per period
$l_t = f(q_t)$	labour time required to produce q_t

The rate of growth of the resource is assumed to be equal to $k \geqslant 0$. Hence, if x_t is left over at the end of period t, there will be $x_t(1+k)$ available during period $t+1$. The non-renewable case is a special case when $k = 0$. The production function $q_t = f(l_t)$ is assumed to be strictly concave and increasing in its argument.

The strictly quasi-concave and twice continuously differentiable utility function of individual i can be written[1]

$$U_i = U_i(q_{i1}, \ldots, q_{iT}, x_{i1}, \ldots, x_{iT}, r_{i1}, \ldots, r_{iT}) \qquad \text{all } i$$

The set of utilityfunctions includes individuals at the time the calculation is carried out as well as members of generations yet unborn.

The intergenerational Pareto optimum problem is stated as

$$\max U_1(\cdot) \tag{10.3}$$

subject to

$$\text{(a)} \quad U_j \geqslant \bar{U}_j \qquad j = 2, \ldots, N$$

$$\text{(b)} \quad \sum_{i=1}^{N} x_{it} + x_t \leqslant (1+k)\, x_{t-1} \qquad \text{all } t$$

$$\text{(c)} \quad \sum_{i=1}^{N} q_{it} + \bar{q}_t \leqslant q_t + \bar{q}_{t-1} \qquad \text{all } t$$

$$\text{(d)} \quad f(q_t) + \sum_{i=1}^{N} r_{it} \leqslant L \qquad \text{all } t$$

[1] This assumption guarantees that a point (x, q, r) that fulfils the first order conditions is a global optimum.

Constraint (a) means that the utility of individual 1 is maximized subject to the utilities of all other individuals in any generation exceeding a given level.[2] Condition (b) means that period t's consumption of X plus what is saved of X cannot exceed what was left from the preceding period added to the growth of the resource. Condition (c) states that consumption of good Q in period t plus what is left for later periods cannot exceed the total output of commodity Q in period t plus what is left from period $t-1$. Finally, (d) means that the time used for work and leisure in every period cannot exceed the total available time in the period.

The optimization problem (10.3) implies that extraction costs are constant (zero) over the planning period. Hence, they are independent of the volume extracted or left over. This is a considerable simplification of the real world. Another lack of realism is involved in the assumption that the planning horizon is finite, but this is less serious since T can be made arbitrarily large.

The Lagrangian of the optimization problem (10.3) is written

$$\begin{aligned} L(q, x, r, \lambda, \mu, \gamma, \Theta) &= U_1(q_{11}, \ldots, q_{1T}, x_{11}, \ldots, x_{1T}, r_{11}, \ldots, r_{1T}) \\ &+ \sum_{j=2}^{N} \lambda_j (U_j - \bar{U}_j) + \sum_{t=1}^{T} \mu_t \left[(1+k) x_{t-1} - x_t - \sum_{i=1}^{N} x_{it} \right] \\ &+ \sum_{t=1}^{T} \gamma_t (\bar{q}_{t-1} + q_t - \bar{q}_t - \sum_{i=1}^{N} q_{it}) \\ &+ \sum_{t=1}^{T} \Theta_t \left[L - f(q_t) - \sum_{i=1}^{N} r_{it} \right] \end{aligned} \tag{10.4}$$

The first order conditions for a maximum are summarized in the appendix at the end of the chapter. We start our discussion of the solution to this problem with a result that follows from the finite planning horizon assumption. The result implies that under certain conditions a renewable resource should be exhausted before the end of the planning horizon:

Proposition 10.1
If the first individual's marginal utility of the resource X in period T is positive and/or the marginal utility of the resource

[2] We sidestep the important but unresolved problem of intergenerational equity. For a discussion of this problem, see Dasgupta and Heal (1979), chapter 9.

is positive for any individual for whom $\lambda_i > 0$, then the resource will be exhausted by the end of period T.

The proof of this proposition is somewhat trivial since it cannot be profitable to have some resource left at the end of the planning horizon if it has a positive marginal utility. The formal proof runs like this. Assume that $\partial U_1/\partial x_{1T} > 0$. Next, differentiate (10.4) with respect to x_T. The Kuhn-Tucker conditions are

$$(1+k)\,\mu_{T+1} - \mu_T \leqslant 0 \qquad x_T[(1+k)\,\mu_{T+1} - \mu_T] = 0 \qquad (10.5)$$

The resource has no value after period T, which means that μ_{T+1} is zero. Moreover, since $\mu_T \geqslant \partial U_1/\partial x_{1T} > 0$ from the first order conditions, it follows from (10.5) that $x_T = 0$. This result does not depend on the fact that X is a (renewable) resource. If $\partial U_i/\partial q_{iT} > 0$, for some i, $\bar{q}_T = 0$, and nothing will be stored beyond the planning horizon.

We now turn to the Pareto optimal pricing problem. We assume that prices are somehow assigned and investigate the values necessary to sustain a Pareto optimal consumption plan. As shown in the appendix at the end of the chapter, for household i that maximizes utility subject to its budget constraint,

$$\begin{aligned} \partial U_i/\partial q_{it} &= v_i p_{1t} \\ \partial U_i/\partial x_{it} &= v_i p_{2t} \qquad \text{all } i \\ \partial U_i/\partial r_{it} &= v_i w_t \end{aligned} \qquad (10.6)$$

i.e. the marginal utility of a commodity equals the present value of the price (multiplied by v_i, the marginal utility of income). Thus prices are discounted to the initial period; the market rate of interest will be taken as a datum throughout the chapter.

By differentiating (10.4) with respect to x_{it} one obtains the first order condition

$$\lambda_i\,\partial U_i/\partial x_{it} - \mu_t \leqslant 0 \qquad x_{it}(\lambda_i\,\partial U_i/\partial x_{it} - \mu_t) = 0 \qquad (10.7)$$

If the household consumes the renewable resource in two consecutive periods (10.5) with T replaced by t, (10.6) and (10.7) give

$$p_{2t}/p_{2t+1} = (\partial U_i/\partial x_{it})/(\partial U_i/\partial x_{it+1}) = \mu_t/\mu_{t+1} = 1 + k \qquad (10.8)$$

Thus, we have:

Proposition 10.2 (Baumol and Oates, 1975)
The present values of the prices that sustain a Pareto optimal allocation in a market economy for a renewable (non-renewable)

resource decline (are constant) at rate $(1 + k)^{-1}$, i.e. at the same rate as the resource reproduces itself.

The common sense explanation for this seemingly paradoxical result is that reduced future prices are essentially a bonus for postponed consumption. The highest bonus/unit that can profitably be paid by society is the growth in volume of the resource between two consecutive periods. If the resource is non-renewable society cannot afford any bonus for the postponement of consumption and the present value of the price of the resource will be constant. This is similar to Hotelling's classical result derived in chapter 2.

One might argue that this result is uninteresting since only the relative price of the natural resource is relevant. For example, from (10.4), if stocks of Q are left in two consecutive periods,

$$p_{1t}/p_{1t+1} = (\partial U_i/\partial q_{it})/(\partial U_i/\partial q_{it+1}) = \gamma_t/\gamma_{t+1} = 1 \qquad (10.9)$$

and hence

$$(p_{2t}/p_{1t})/(p_{2t+1}/p_{1t+1}) = p_{2t}/p_{2t+1} = 1 + k$$

which means that the present value of the price of the resource expressed in terms of the present value of the price of the storable commodity Q will also decline at rate $1 + k$.

10.4 Optimal Pricing Rules when there are Several Regenerating Resources

Following Baumol and Oates, we now introduce the possibility that Q might grow at a constant rate $g > 0$. In other words, a stock of Q that is left to a later period will be augmented to $(1 + g)\bar{q}_t$. Q might for example be timber which, once planted with labour, grows at the rate g. A forest does not grow exponentially. Hence the assumption that it does involves considerable unrealism.[3] Nevertheless, we believe that the conclusions to follow on necessary conditions for a Pareto optimal price system tell us something important about the optimal pricing of different kinds of natural resources.

The balance condition (10.3c) now reads

$$\sum_{i=1}^{N} q_{it} + \bar{q}_t \leqslant q_t + (1 + g)\bar{q}_{t-1}$$

[3] See also section 11.2, where we explicitly distinguish between different kinds of renewable resources.

The Kuhn-Tucker conditions obtained when differentiating (10.4) with respect to $\bar{q}_t$ are

$$(1+g)\,\gamma_{t+1}-\gamma_t \leqslant 0 \qquad \bar{q}_t[(1+g)\,\gamma_{t+1}-\gamma_t]=0 \tag{10.10}$$

If Q and X are consumed in two consecutive periods, the following conditions are necessary for Pareto optimal relative prices [provided that (10.10) holds with equality, which is the case if $\bar{q}_t > 0$]:

$$\begin{aligned} p_{1t}/p_{1t+1} &= 1+g \\ p_{2t}/p_{2t+1} &= 1+k \\ (p_{2t}/p_{1t})/(p_{2t+1}/p_{1t+1}) &= (1+k)/(1+g) \end{aligned} \tag{10.11}$$

In short:

> **Proposition 10.3** (Baumol and Oates, 1975)
> The prices that sustain a Pareto optimum for a resource that regenerates itself at rate $k > 0$ ($g > 0$) will always fall with time. However, if expressed in terms of some commodity that grows in excess of k, the current price of the resource will rise with time.

In other words, under a Pareto optimal price system, the relative (with respect to say timber) price of a resource that regenerates itself extremely slowly should increase over time. Intuitively, although X grows, Q might grow even faster. This means that society can pay a higher bonus for the postponement of the consumption of Q than of X. In other words, the optimal pricing system should encourage the early consumption of X and this is done by a price system that makes Q more expensive, in terms of X, in later periods.

If the commodities are *perfect substitutes* in consumption, i.e. the commodities give the same additions to utility, or formally

$$\begin{aligned} &U_i(q_{i1},\ldots,q_{iT},x_{i1},\ldots,x_{iT},\ldots) \\ &\quad = U_i(x_{i1}+q_{i1},x_{i2}+q_{i2},\ldots) \qquad \text{all } i \end{aligned}$$

then the resource that grows the slower should be consumed first. In fact, one might expect that the consumption of the fast growing producible resource Q should be postponed until the slow growing non-producible resource X is exhausted. This can be proved:

> **Proposition 10.4** (Baumol and Oates, 1975)
> If X and Q are perfect substitutes in consumption, but it is optimal for any person to consume commodity Q during some

t, it will never be optimal for anyone to consume X in any period after t (if $0 \leqslant k < g$).

Proof

Assume the contrary, i.e. that $q_{jt} > 0$ and $x_{it+1} > 0$ for some individual is i, j. Since the commodities by assumption are perfect substitutes, it follows from (10.4) by differentiation with respect to x_{it}, x_{it+1}, q_{jt} and q_{jt+1} that

$$\mu_t \geqslant \lambda_i \partial U_i / \partial x_{it} = \lambda_j \partial U_j / \partial q_{jt} = \gamma_t$$

and

$$\mu_{t+1} = \lambda_i \partial U_i / \partial x_{it+1} = \lambda_j \partial U_j / \partial q_{jt+1} \leqslant \gamma_{t+1} \tag{A}$$

Moreover, since $x_{it+1} > 0$ and X is not producible, it must hold that $x_t > 0$. Hence, differentiation of (10.4) with respect to x_t and $\bar{q}_t$ gives

$$(1 + k)\, \mu_{t+1} = \mu_t \geqslant \gamma_t \geqslant \gamma_{t+1}(1 + g)$$

and since $0 \leqslant k < g$ by assumption, one has

$$\mu_{t+1} > \lambda_{t+1} \tag{B}$$

which contradicts (A). The claim that $q_{jt} > 0$ implies that $x_{it+1} = 0$ for all i has thereby been proved, and it is straightforward to extend the argument to any future period.

Loosely speaking, if it is possible to produce fuels from oil and timber that are perfect substitutes, in a Pareto optimal programme one would first use up the oil, and then start to produce fuel from timber. However, several qualifications hold. First, oil and timber are not perfect substitutes. Second, timber does not grow exponentially and without limit – in fact it might shrink for t larger than some t^*.

Finally, oil and (or) fuel from oil cannot be produced costlessly. It should, however, be pointed out that the fact that growth is assumed to be exponential is unnecessarily restrictive. This result will of course hold under more general growth assumptions. For example, if the growth rates (g and k) are functions of the stocks left for later periods,

$$k_t = k(x_t) \geqslant 0 \qquad \text{all } x_t \geqslant 0$$

$$g_t = g(\bar{q}_t) \geqslant 0 \qquad \text{all } \bar{q}_t \geqslant 0$$

and all previous conclusions in principle hold. Note, however, that the marginal growth rates k_{m} and g_{m} are relevant, not the average growth rates. More specifically, the proof of proposition 10.4 holds if it is assumed that $g_m > k_m$ for all $\bar{q}$ and x. If Q is assumed to be timber, it is well known that one might have $g_m < 0$ for some $\bar{q}^* > \bar{q}$, but $g_m < 0$ can easily be shown to imply that nothing is left for later periods, and cannot hold in the interior of an optimal programme. Note that $g_m < 0$ does not mean that what is left over is not augmented $[g(\bar{q}_t) < 0]$. It only means that one cannot further increase the volume of the resource in the next period by leaving another unit to grow.

The almost trivial proposition 10.1 is independent of the growth functions g and k, while both propositions 10.2 and 10.3 become a lot more diffuse and less specific by the introduction of more general growth functions.

10.5 Multiple Use Considerations

In this section, we introduce multiple use considerations into the welfare problem. This is done by assuming that the stock of a resource, say Q, has a value to consumers. For example, if Q is timber a forest may have recreation value. Formally, we rewrite the utility functions as

$$U_i(q_{i1}, \ldots, r_{iT}, \bar{q}_1, \ldots, \bar{q}_T) \qquad \text{all } i \qquad (10.12)$$

The problem of finding an intergenerational Pareto optimum can then be stated as in (10.3), but with the growth rate g strictly positive to keep the problem as general as possible (see section 10.4).

We start our discussion of multiple use considerations with a proof of the following claim,[4] which is also illustrated in figure 10.2:

Proposition 10.5
The resource Q will be exhausted at the end of period T in a Pareto optimum if the marginal utility of the consumption of the resource in period T is positive for any individual (for whom $\lambda_j > 0$) and larger than the weighted sum of all individuals' marginal utilities of the stock of the resource at the end of period T.

[4] The reader could, if he/she wishes, derive the results that follow from the more general growth function $g(\bar{q}_t)$, and note that $g_m \leqslant 0$ and $\bar{q} > 0$ are now consistent

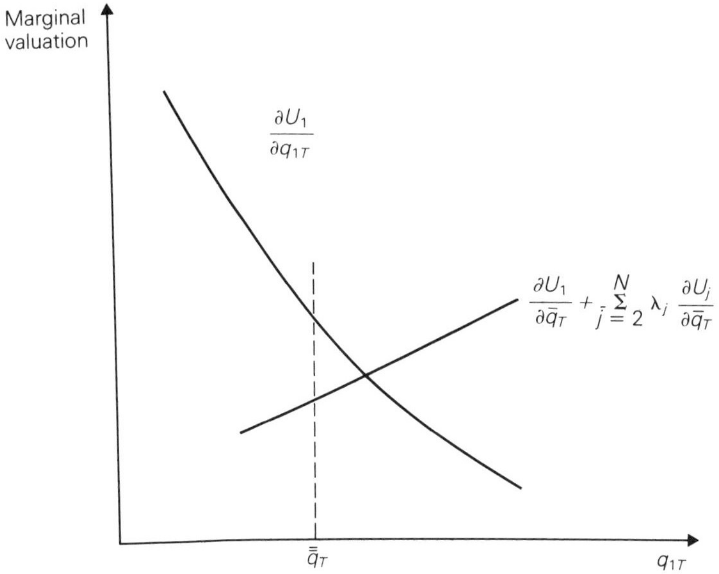

FIGURE 10.2 Suppose there remain $\bar{\bar{q}}_T$ units of the resource in period T. Then the resource will be exhausted ($q_{1T} = \bar{\bar{q}}_T$) since the first individual's marginal utility of consuming the resource exceeds the weighted sum of each individual's marginal utility of having the last unit of the resource saved

Proof

To prove this, it suffices to differentiate (10.4) with respect to $\bar{q}_T$, using the 'new' utility functions (10.12):

$$\sum_{i=1}^{N} \lambda_i (\partial U_i / \partial \bar{q}_T) - \gamma_T \leqslant 0 \tag{10.13}$$

$$\bar{q}_T \sum_{i=1}^{N} [\lambda_i (\partial U_i / \partial \bar{q}_T) - \gamma_T] = 0$$

and $\bar{q}_T = 0$ if the first equation holds with inequality. Note that γ_{T+1} has been set equal to zero since the planning horizon covers only the first T periods. The result then follows since $\Sigma \lambda_j (\partial U_j / \partial \bar{q}_T)$ is the weighted sum of all individuals' marginal utilities of the stock of the resource and $\lambda_j \partial U_j / \partial q_{jT} \leqslant \gamma_T$. The latter inequality, which holds as an equality if $q_{jT} > 0$, is obtained by differentiation of (10.4) with respect to q_{jT}.

The reader might also note the trivial result that a resource whose stock has a negative marginal utility to all individuals at the end of period T will always be exhausted.

We now turn to the *Pareto sustaining* price system. For an individual who consumes the resource Q in two consecutive periods, a price system that is consistent with the requirements of Pareto optimum must fulfil

$$p_{1t}/p_{1t+1} = (\partial U_i/\partial q_{it})/(\partial U_i/\partial q_{it+1}) = \gamma_t/\gamma_{t+1} \tag{10.14}$$

and, provided that $\bar{q}_t > 0$,

$$0 < p_{1t}/p_{1t+1} = (1+g)/[1-(1/\gamma_t)\,\Sigma\lambda_i\,\partial U_i/\partial\bar{q}_t] \geqslant 1+g \tag{10.15}$$

since

$$\Sigma\lambda_i(\partial U_i/\partial\bar{q}_t) + (1+g)\,\gamma_{t+1} - \gamma_t = 0 \qquad \text{if } \bar{q}_t > 0$$

Thus:

Proposition 10.6
The present values of prices that sustain a Pareto optimal allocation in a market economy for a renewable (non-renewable) resource, whose stock has a positive weighted sum of marginal utilities, declines faster than the constant rate at which the resource reproduces itself.

If we explicitly introduce the other resource X into the model, whose stock is not valued, it should be fairly obvious that it is possible to prove the following proposition:

Proposition 10.7
The prices that sustain a Pareto optimum for a renewable or non-renewable resource, whose stock has a positive weighted sum of marginal utilities, always decline over time. If expressed in terms of some commodity that grows at a constant rate in excess of g, prices might still decline over time. If the *numéraire* resource grows at constant rate k, the relative price of the resource that has value as a stock declines if

$$\frac{1+g}{1+k} > 1 - \frac{1}{\gamma_t}\,\Sigma\lambda_i(\partial U_i/\partial\bar{q}_t)$$

The intuitive explanation for proposition 10.6 is that a positive stock valuation introduces a bonus in excess of that created through growth. This extra bonus might outweigh the fact that the *numéraire* resource grows faster than the multiple use resource, explaining proposition 10.7.

We end this chapter by pointing out that proposition 10.4 can be proven if the stock of Q is valued and grows faster than another non-producible resource X. This is guaranteed by the fact that the resources are assumed to be perfect substitutes. It should also be emphasized that the results derived in this chapter are unchanged if additional consumption goods are introduced into the utility functions. The number of goods has been kept as low as possible to simplify the exposition. The reader should also note that in section 11.4 we will return to the multiple use issue. There we will provide a more explicit discussion and model non-timber uses of forest lands.

Appendix: Complete First Order Conditions for some of the Maximization Problems

The Lagrangian of the optimization problem (10.3) was written

$$\begin{aligned} &L(q, x, r, \lambda, \mu, \gamma, \Theta) \\ &= U_1(q_{11}, \ldots, q_{1T}, x_{11}, \ldots, x_{1T}, r_{11}, \ldots, r_{1T}) \\ &\quad + \sum_{j=2}^{N} \lambda_j (U_j - \bar{U}_j) + \sum_{t=1}^{T} \mu_t \left[(1+k) x_{t-1} - x_t - \sum_{i=1}^{N} x_{it} \right] \\ &\quad + \sum_{t=1}^{T} \gamma_t \left(\bar{q}_{t-1} + q_t - \bar{q}_t - \sum_{i=1}^{N} q_{it} \right) \\ &\quad + \sum_{t=1}^{T} \Theta_t \left[L - f(q_t) - \sum_{i=1}^{N} r_{it} \right] \end{aligned} \qquad (10.4)$$

The first-order conditions for maximum can be written:

(a) $U_{1x_t} - \mu_t \leqslant 0 \quad x_{1t}(U_{1x_t} - \mu_t) = 0 \quad \text{all } t$

(b) $\lambda_j U_{jx_t} - \mu_t \leqslant 0 \quad x_{jt}(\lambda_j U_{jx_t} - \mu_t) = 0 \quad \text{all } t \quad j - 2, \ldots, N$

(c) $U_{1q_t} - \gamma_t \leqslant 0 \quad q_{1t}(U_{1q_t} - \gamma_t) = 0 \quad \text{all } t$

(d) $\lambda_j U_{jq_t} - \gamma_t \leqslant 0 \quad q_{jt}(\lambda_j U_{jq_t} - \gamma_t) = 0 \quad \text{all } t \quad j = 2, \ldots, N$

(e) $U_{1r_t} - \Theta_t \leqslant 0 \quad r_{1t}(U_{1r_t} - \Theta_t) = 0 \quad \text{all } t$

(f) $\lambda_j U_{jr_t} - \Theta_t \leqslant 0 \quad r_{jt}(\lambda_j U_{jr_t} - \Theta_t) = 0 \quad \text{all } t \quad j = 2, \ldots, N$

(g) $U_j - \bar{U}_j \geqslant 0 \quad \lambda_j(U_j - \bar{U}_j) = 0 \quad j = 2, \ldots, N$

(h) $(1+k)x_{t-1}-x_t-\sum_{i=1}^{N} x_{it} \geqslant 0$

$$\mu_t\left[(1+k)x_{t-1}-x_t-\sum_{i=1}^{N} x_{it}\right]=0 \quad \text{all } t \tag{10.16}$$

(i) $\bar{q}_{t-1}+q_t-\bar{q}_t-\sum_{i=1}^{N} q_{it} \geqslant 0$

$$\gamma_t\left(\bar{q}_{t-1}+q_t-\bar{q}_t-\sum_{i=1}^{N} q_{it}\right)=0 \quad \text{all } t$$

(j) $L-f(q_t)-\sum_{i=1}^{N} r_{it} \geqslant 0 \quad \Theta_t\left[L-f(q_t)-\sum_{i=1}^{N} r_{it}\right]=0 \quad \text{all } t$

(k) $(1+k)\mu_{t+1}-\mu_t \leqslant 0 \quad x_t[(1+k)\mu_{t+1}-\mu_t]=0 \quad \text{all } t$

(l) $\gamma_{t+1}-\gamma_t \leqslant 0 \quad \bar{q}_t(\gamma_{t+1}-\gamma_t)=0 \quad \text{all } t$

(m) $\gamma_t-\Theta_t f'(q_t) \leqslant 0 \quad q_t[\gamma_t-\Theta_t f'(q_t)]=0 \quad \text{all } t$

where

$$U_{ix_t}=\frac{\partial U_i}{\partial x_{it}}$$

$$U_{iq_t}=\frac{\partial U_i}{\partial q_{it}}$$

$$U_{ir_t}=\frac{\partial U_i}{\partial r_{it}}$$

$$f'(q_t)=\frac{\partial f}{\partial q_t}$$

These are $3NT+N-1+3T+3T$ equations in the same number of unknowns, and the parameters that determine the optimal solution are the initial stock of the resource x_0, the rate of growth of the resource k, and the total labour time available per period L.

To arrive at the first order conditions (10.6) for utility maximization, we introduce the following 'lifetime' budget constraint of the ith household:

$$\pi_i+\sum_{t=1}^{T}[w_t(L-r_{it})-p_{1t}q_{it}-p_{2t}x_{it}]=0 \quad \text{all } i \tag{10.17}$$

where π_i is the present value of a lump sum income, w_t is the present value of the wage rate in period t, p_{1t} is the present value of the price of Q and p_{2t} is the present value of the price of X.

The household uses its total lump sum plus labour income to consume goods.

The Lagrangian of the relevant optimization problem is written

$$H_i(q, x, r, v_i) = U_i(q_{i1}, \ldots, r_{iT}) + v_i \left\{ \pi_i + \sum_{t=1}^{T} [w_t(L - r_{it}) - p_{1t}q_{it} - p_{2t}x_{it}] \right\} \quad \text{all } i \tag{10.18}$$

The first order conditions for an interior maximum are

$$\begin{aligned} \frac{\partial H_i}{\partial q_{it}} &= \frac{\partial U_i}{\partial q_{it}} - v_i p_{1t} = 0 \\ \frac{\partial H_i}{\partial x_{it}} &= \frac{\partial U_i}{\partial x_{it}} - v_i p_{2t} = 0 \qquad \text{all } t \\ \frac{\partial H_i}{\partial r_{it}} &= \frac{\partial U_i}{\partial r_{it}} - v_i w_t = 0 \\ \frac{\partial H_i}{\partial v_i} &= \pi_i + \sum_{t=1}^{T} [w_t(L - r_{it}) - p_{1t}q_{it} - p_{2t}x_{it}] = 0 \end{aligned} \tag{10.19}$$

These are the conditions used in (10.6).

For a further analysis of the behaviour of households (and firms) in a market economy, the reader is referred to chapter 11.

11

Cost–Benefit Rules for Natural Resources

11.1 Introduction

Chapter 10 was devoted to an examination of the properties of a Pareto sustaining price system. A price vector is Pareto sustaining if it ensures an allocation such that production and distribution cannot be reorganized to increase the utility of one individual without decreasing the utility of at least one other individual. In real world economies, the actual price vector may be different from the Pareto sustaining vector. Suppose, however, that we can estimate a Pareto sustaining price vector for say the United States. Then it is possible to design a set of taxes and subsidies which moves the US economy to a Pareto optimal state. Furthermore, using these prices it can be determined whether or not a certain project, say increased yearly harvest of a renewable resource, is socially desirable. This approach, however, requires such huge amounts of information that it is generally inoperable.

In this chapter we turn to an alternative approach which is more realistic since it demands less information. We use a social cost–benefit analysis to determine the small movements in production and employment which improve welfare. Starting from a situation which is not Pareto optimal, a (social) cost–benefit analysis may be used to examine whether a certain project (or a tax or a subsidy) leads to a Pareto improvement.

To derive cost–benefit rules for an economy which uses non-renewable and renewable natural resources, we develop an inter-temporal model of an open economy. The country considered has a single industrial sector producing a homogeneous and traded good, say pulp, using timber and labour as variable inputs.[1] Different kinds

[1] There is no real loss of generality and the exposition is simplified if it is assumed that the country has a single industrial sector (the forest industry).

of natural resources, which are also traded internationally, are extracted or harvested using labour as the sole variable input. Households consume commodities and supply labour. To generate appropriate cost–benefit rules, we follow the tradition in welfare economics of considering state owned firms that produce marketable goods – either ordinary produced goods or different kinds of natural resources. Their levels of production (extraction) are exogenous policy determined variables. Hence these firms do not necessarily maximize profits. They are, however, price takers like all other agents in the economy.

The chapter is structured in the following way. Section 11.2 develops the model for the case where there are no market imperfections. Appropriate cost–benefit rules for this case are presented in section 11.3. In particular, it is demonstrated that the traditional cost–benefit rules found in textbooks on welfare economics in general are incorrect or at least misleading when applied to non-renewable or renewable natural resources.[2] Section 11.4 introduces a particular kind of problem, namely multiple uses of natural resources, and discusses the treatment of such a complication in project evaluations. Then in sections 11.5 and 11.6 we address the empirically important question of whether the socially optimal rotation differs from the Faustmann rotation when the economy suffers from market imbalances like unemployment and excess demand for timber. Section 11.7 contains a few remarks about the future use of cost–benefit analysis in forest economics. An alternative approach (maximization of a so-called indirect utility function) to that used in the main text is discussed in the appendix at the end of the chapter.

11.2 The Model

This section considers a small open economy which can buy and sell without limit in each period at fixed foreign currency prices. Given a fixed exchange rate, and assuming the law of one price holds, the domestic currency prices of both produced goods and natural resources are also fixed. The country will be a net exporter or importer of any traded good depending on the level of the fixed world market price of the good. The argument is illustrated in figure 11.1a, where the price is fixed at such a level that the country is a

[2] See Nautiyal and Rezende (1983) for a discussion of dynamic investment considerations in forestry project evaluations. See also Dasgupta (1982) and Johansson (1983; 1984a, b, c).

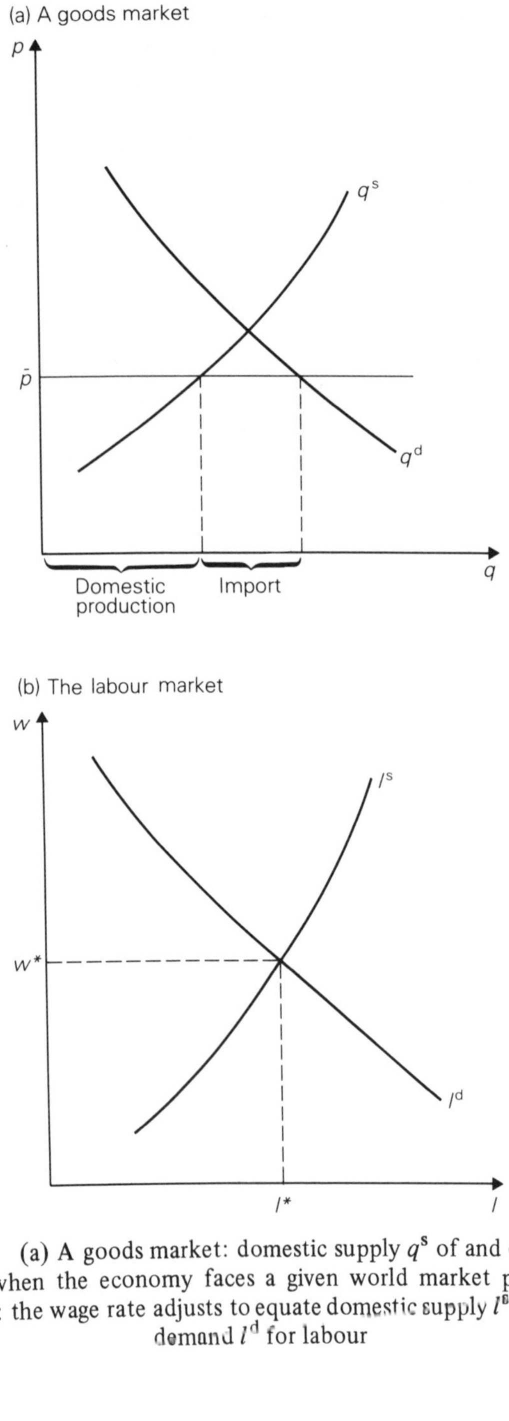

FIGURE 11.1 (a) A goods market: domestic supply q^s of and demand q^d for a traded good when the economy faces a given world market price $\bar{p}$. (b) The labour market: the wage rate adjusts to equate domestic supply l^s of and domestic demand l^d for labour

net importer of the good. Hence the balance of trade, i.e. the value of exports less the value of imports, need not be in equilibrium in each and every year. This will be discussed in section 11.3. Moreover, assuming perfect capital mobility and a single internationally traded bond, the interest rate (the rate of return) is exogenously fixed at the world level for the small open economy under consideration. The nominal wage rate is allowed to adjust in each period so as to achieve full employment (see figure 11.1b).

11.2.1 *Private Producers*

The forest industry. Within this environment, the forest industry produces a homogeneous and non-storable product q_{1t} with timber c_{1t} and labour l_{1t} as variable inputs. The industry maximizes the sum of its profits during T periods discounted to the initial period:

$$\pi_1 = \sum_{t=1}^{T} [p_{1t} f_{1t}(c_{1t}, l_{1t}) - p_t c_{1t} - w_t l_{1t}] \tag{11.1}$$

where subscript 1 denotes the forest industry, p_{1t} is the present value of the price of the forest industry's output in period t, p_t is the present value of the timber price in period t, w_t is the present value of the wage in period t, f_{1t} is the strictly concave and twice continuously differentiable production function of the forest industry in period t, c_{1t} is the forest industry's demand for timber in period t and l_{1t} is the forest industry's demand for labour in period t. Any fixed costs will be ignored throughout the chapter since such costs do not affect the marginal cost-benefit rules to be derived (see section 11.3).

The necessary (and sufficient) condition for profit maximization is that the value of the marginal product of each factor in the tth period is equal to its price in that period:

$$\left.\begin{aligned} p_{1t}\frac{\partial f_{1t}}{\partial c_{1t}} - p_t &= 0 \\ p_{1t}\frac{\partial f_{1t}}{\partial l_{1t}} - w_t &= 0 \end{aligned}\right\} \quad \text{all } t \tag{11.2}$$

In (11.2) prices and wages are discounted to the initial period, but we could just as well have expressed them in current terms. The current prices are obtained by multiplying by $(1 + r)^{t-1}$, where r is the world market determined rate of interest (which for simplicity is assumed to be the same in all periods).

Firms extracting the non-renewable resource. Private sector firms extracting the non-renewable resource q_{2t} are assumed to use labour as the only variable production factor. The representative firm maximizes

$$L_2 = \sum_{t=1}^{T} [p_{2t} f_{2t}(l_{2t}) - w_t l_{2t}] + \lambda_2 \left[\bar{x}_2 - \sum_{t=1}^{T} f_{2t}(l_{2t}) \right] \quad (11.3)$$

where subscript 2 is used to indicate that the firm extracts a non-renewable resource, and f_{2t} is the strictly concave and twice continuously differentiable production function. The firm cannot extract more than the initial stock $\bar{x}_2$ of the resource. This explains that there is a constraint on the maximization problem over and above the constraint (i.e. the production technology) faced by the forest industry in (11.1).

An interior solution to (11.3) requires that

$$\partial L_2/\partial l_{2t} = p_{2t} f'_{2t} - w_t - \lambda_2 f'_{2t} = 0 \quad \text{all } t \quad (11.4)$$

where a prime is used to denote a partial derivative with respect to l_{2t}. According to (11.4), in each period of extraction there is a positive *difference* between the value of the marginal product and the wage if the resource constraint is binding. In fact, the present value of the marginal profit is constant and equal to $\lambda_2 > 0$ in all periods of extraction (and not equal to zero as it was for producers of ordinary goods). The Lagrange multiplier λ_2 may be interpreted as the present value of future profits lost if extraction today is increased by one unit (or as the increase in the present value of profits if the stock of the resource increases by one unit).

Note, however, that if there are no costs of extraction ($w_t l_{2t} = 0$ all t), then we have Hotelling's (1931) well-known rule (derived in chapter 2) that extraction will take place in two consecutive periods if the discounted price of the resource is constant over time:

$$p_{2t} = \lambda_2 = p_{2t+1} \quad (11.5a)$$

i.e. if the rate of change of the world market determined price in *current terms* happens to equal the interest rate:

$$p^c_{2t+1} = p^c_{2t}(1 + r) \quad (11.5b)$$

where a superscript c denotes a price in current terms. If the price grows at a faster rate, it pays to keep the stock intact and accumulate capitals gains. A slower rate of change of the price than the interest rate induces the firm to immediately extract the resource. Recall that firms are assumed to be able to sell whatever quantities they wish at the prevailing world market price. For a further discussion of extraction rules for non-renewable resources, the reader is referred to chapter 2. Note, however, that a small open economy cannot affect world market prices. In the *closed* economy model considered in chapter 2, on the other hand, the domestic price of the resource adjusts to equate domestic supply and domestic demand.

Firms harvesting a renewable resource other than timber. For the moment, we want to avoid the so-called problem of the commons dealt with in section 3.3. Hence it is assumed that firms harvesting the renewable resource q_{3t} own the resource. These firms use labour as the only variable input, but can also 'select' the stock of the resource by harvesting more or less than the natural growth of the resource. The representative firm maximizes

$$L_3 = \sum_{t=1}^{T} [p_{3t}f_{3t}(l_{3t}) - w_t l_{3t}] + \sum_{t=1}^{T} \lambda_{3t}[x_{3t-1} + F_t(x_{3t-1}) - f_{3t}(l_{3t}) - x_{3t}] \tag{11.6}$$

where subscript 3 is used to indicate that the firm harvests a renewable resource, $f_{3t}(l_{3t})$ is the strictly concave and twice continuously differentiable production function in period t, $F_t(x_{3t-1})$ represents the growth of the resource in period t as a function of the stock at the end of period $t-1$ with $F_t' \gtreqless 0$ and $F_t'' \leqslant 0$.

In (11.6) there is a constraint on the harvest such that the sum of the harvest in period t plus what is left to period $t+1$ cannot exceed what was left over from period $t-1$ plus the natural growth of the resource in period t. From this formulation it should be obvious that the non-renewable case treated above is a special case when there is no growth of the resource. The necessary conditions for profit maximization (if we restrict ourselves to interior solutions) are

$$\partial L_3/\partial l_{3t} = p_{3t}f_{3t}' - w_t - \lambda_{3t}f_{3t}' = 0 \tag{11.7}$$

$$\partial L_3/\partial x_t = -\lambda_{3t} + \lambda_{3t+1}(1 + F_{t+1}') = 0 \tag{11.8}$$

$$\partial L_3/\lambda_{3t} = x_{3t-1} + F_t(x_{3t-1}) - f_{3t}(l_{3t}) - x_{3t} = 0 \tag{11.9}$$

To interpret these conditions, we focus on a steady state solution, i.e. where the firm harvests the growth in each period so that the stock of the resource remains constant over time. Using conditions (11.7) and (11.8) respectively for two consecutive periods yields

$$\lambda_{3t} = \lambda_{3t+1}(1 + r) \tag{11.7a}$$

$$\lambda_{3t} = \lambda_{3t+1}(1 + F') \tag{11.8a}$$

To arrive at these expressions, we have used the fact that the harvest is constant from period to period in a steady state (so that $f'_{3t} = f'_{3t+1}$). Furthermore, it is assumed that all prices are constant in current terms, i.e. the discounted prices decrease over time: $p_{3t} = p_{3t+1}(1+r)$, $w_t = w_{t+1}(1+r)$. Conditions (11.7a) and (11.8a) imply that a profit maximizing firm selects a stock such that the growth (F') equals the market rate of interest (r). This is exactly the result obtained in (3.8).

In order to maximize its total profits over a horizon containing T periods, the firm must liquidate the (steady state) stock of the resource at the end of period T. Owing to the small open economy assumption, the firm can sell the property rights to the stock in the world market. However, as T tends to infinity, the present value of this revenue tends to zero. This is why the terminal value of the stock discounted to the initial period is not included in (11.6). The same argument, of course, applies to other resources, like forest lands which are owned by domestic residents.

Private timber producers. Private timber producers, operating on a fixed land base, are assumed to use labour as the only variable input when planting and cutting trees. Assume that current values of the timber price p_t^c, the wage w_t^c and the interest rate r are expected to be constant over time. The optimum or steady state rotation T^* for a fully regulated forest then results in a harvest level c^* at each cutting. The present value of profits of the 'representative' timber producer is

$$\pi_4 = (p_1 c_1^* - w_1 l_{41}) + (p_2 c_2^* - w_2 l_{42}) + \dots \\ + (p_1 c_1^* - w_1 l_{41})(1+r)^{-T^*} + (p_2 c_2^* - w_2 l_{42})(1+r)^{-T^*} + \dots \tag{11.10}$$

where $p_t = p_1(1+r)^{-t+1}$, $w_t = w_1(1+r)^{-t+1}$, c_1^* represents trees planted T^* years ago, c_2^* trees reaching the age of T^* years in period 2, and so on. Once a tree is cut the area is replanted and the tree is again allowed to grow to age T^*, etc. For simplicity, harvesting costs $w_t l_{4t}$ (including replanting costs) are assumed to be proportional to the harvest level: $w_t l_{4t} = w_t(ac_t)$, where a is a constant. This implies that (11.10) could be simplified by defining a timber price net of cutting costs $(p_t - aw_t)$, but in deriving cost-benefit rules it will prove useful to consider cutting costs explicitly, as is done in (11.10).

To highlight the nature of the profit maximizing solution, we consider a single tree or an even-aged stand that reaches the age of T^* years in period 2. The present value of profits from all future rotations is

$$\begin{aligned}\hat{\pi}_4 &= (p_2 c_2^* - w_2 l_{42})\,[1 + (1+r)^{-T^*} + (1+r)^{-2T^*} + \ldots] \\ &= (p_2 c_2^* - w_2 l_{42})/[1 - (1+r)^{-T^*}] \qquad (11.11)\end{aligned}$$

By allowing the time horizon T to tend to infinity, the number of rotation periods (T/T^*) will also tend to infinity. This assumption has been employed in arriving at the second equality in (11.11).

Next, suppose the firm shortens all rotations by one 'year' (one period), i.e. trees are now cut when they reach an age of $T^* - 1$ years. The present value of all future profits will now be

$$\begin{aligned}\tilde{\pi}_4 &= (p_1 c_2^* - w_1 l_{42})(1+K)^{-1}[1 + (1+r)^{-T^*+1} + (1+r)^{-2T^*+2} + \ldots \\ &= (p_1 c_2^* - w_1 l_{42})(1+K)^{-1}/[1 - (1+r)^{-T^*+1}] \qquad (11.12)\end{aligned}$$

where K is the annual growth rate of trees when they have an age of $T^* - 1$ years, i.e. $c_2^* = (1 + K)\,c_1$ or $c_1 = c_2^*(1 + K)^{-1}$.

From the derivation of the Faustmann rotation in section 4.2 it should be clear that a marginal change in the optimal rotation period leaves the present value of all future profits unchanged ($\mathrm{d}\pi_4/\mathrm{d}T^* = \mathrm{d}V/\mathrm{d}T^* = 0$). A careful examination of (11.11) and (11.12) reveals that $\hat{\pi}_4 = \tilde{\pi}_4$ *if* the marginal growth rate is given by

$$K^* = r + r/[(1+r)^{T^*-1} - 1] \qquad (11.13\text{a})$$

or

$$K^*(p_1 c_1 - w_1 l_{41}) = r(p_1 c_1 - w_1 l_{41}) + r\tilde{\pi}_4(1+r)^{-T^*+1} \qquad (11.13\text{b})$$

This is a discrete time version of the Faustmann optimality condition. A tree ought to be cut when its annual growth rate is equal to the interest on the profit of the cut plus interest on the value of the forest land. (The value of the forest land is defined as the present value of profits in year $T^* - 1$, year $2T^* - 2$, etc.). If the actual growth rate K exceeds (falls short of) K^* it pays to lengthen (shorten) the rotation age. The reader should note that growth rate K (assumed to be solely a function of time) is exogenous in the model since it is rather difficult to model K as a function of time in a discrete time model. Hence, we assume that $K = K^*$ when a tree has an age of $T^* - 1$ years, and that $K_t > K^*$ ($K_t < K^*$) for $t < T^* - 1$ ($t \geqslant T^*$), where t denotes the actual age of a tree.

11.2.2 *The Public Sector*

To generate cost–benefit rules, we follow the tradition in welfare economics and introduce state owned firms. Because the levels of public sector output are exogenously determined policy variables, the behaviour of these firms need not reflect profit maximization.

Any profits (or losses) incurred by state owned firms are assumed to be disposed of (financed) by lump sum taxes (transfers) π_g. Hence the government budget constrain takes the form

$$\begin{aligned}\pi_g &= \sum_{i=1}^{4} \pi_{gi} \\ &= \sum_{t=1}^{T} \{[p_{1t}f_{g1t}(c_{g1t}, l_{g1t}) - p_t c_{g1t} - w_t l_{g1t}] && \text{(A)} \\ &\quad + [p_{2t}f_{g2t}(l_{g2t}) - w_t l_{g2t}] && \text{(B)} \\ &\quad + [p_{3t}f_{g3t}(l_{g3t}) - w_t l_{g3t}]\} && \text{(C)} \\ &\quad + (p_2 c_{g2} - w_2 l_{g42})/[1-(1+r)^{-T_g}] && \text{(D)}\end{aligned} \tag{11.14}$$

where $\pi_g > 0$ is the pesent value of lump sum transfer payments to households and $\pi_g < 0$ the present value of lump sum taxes.

According to (11.14), there is a state-owned firm (A) producing the industrial good (pulp) using timber and labour as variable inputs. Since its level of production is a policy instrument, the values of the marginal product of timber and labour need not equal the timber price p_t and the wage w_t, respectively, which was the case for the profit maximizing firm in (11.2).

Extraction of the non-renewable resource (B) by the public sector need not reflect profit maximization either. However, the state owned firm, like private sector firms in (11.3), cannot extract more than its initial stock $\bar{x}_{g2}$ of the resource.

A similar restriction is faced by the firm (C) harvesting a renewable resource (other than timber). In period t it can harvest no more than what was left over from period $t-1$ plus the natural growth of the resource in period t minus what is planned to be left over to future periods $[f_{g3t}(l_{g3t}) \leqslant x_{gt-1} + F_{gt}(x_{gt-1}) - x_{gt}]$.

Finally, managers of national forests (D) are assumed to select a uniform, but not necessarily profit maximizing, rotation T_g. To simplify notation and exposition we consider only a single tree or an even-aged stand (which reaches the age of T_g years in period 2, period $T_g + 1$, and so on). For simplicity, cutting costs are assumed to be proportional to the volume harvested: $w_2 l_{g42} = w_2(bc_{g2})$, where b is a constant.

As stressed earlier, public sector firms do not necessarily maximize profits. However, it is sometimes assumed that they, like private sector firms, do maximize profits. If so, the conditions for profit maximization derived in section 11.2.1 are assumed to also hold for the state owned firms. We just add a subscript g to highlight that the firm considered is state owned.

11.2.3 *Households*

In order to focus on efficiency considerations, while setting aside matters of (intergenerational) equity and income distribution, the commonly employed assumption of a 'representative' household is used. This representative household consumes all kinds of commodities, supplies labour, and maximizes the strictly quasi-concave and twice continuously differentiable utility function[3]

$$U(q_{11}^{d}, \ldots, q_{3T}^{d}, c_1^{d}, \ldots, c_T^{d}, l_1^{s}, \ldots, l_T^{s}) \tag{11.15}$$

subject to its lifetime budget constraint

$$\sum_{i=1}^{4} (\pi_i + \pi_{gi}) + \sum_{t=1}^{T} \left(w_t l_t^{s} - \sum_{i=1}^{3} p_{it} q_{it}^{d} - p_t c_t^{d} \right) = 0 \tag{11.16}$$

Both borrowing and lending in any amount are allowed at the prevailing interest rate. In addition, we assume that the household plans to leave its 'heirs' neither assets nor debts. Hence the present value of the budget sums to zero. Furthermore, all profits from privately owned firms and state owned firms are distributed to the household. [$\pi_1(\pi_{g1})$ are profits from privately owned (state owned) firms producing pulp, $\pi_2(\pi_{g2})$ are profits from firms extracting the non-renewable resource, $\pi_3(\pi_{g3})$ are profits from firms harvesting a renewable resource other than timber, and $\pi_4(\pi_{g4})$ are profits from timber producers. The term *profits* is used to denote the present value of the sum of profits in all periods.]

Maximization of (11.15) subject to (11.16) yields the following first order conditions for utility maximization (provided the solution is interior as is assumed throughout):

[3] The utility function (11.15) can alternatively be interpreted as a social welfare function. For an informed discussion see Dasgupta and Heal (1979), chapter 9. The multihousehold case will be considered in section 11.4.3.

$$\begin{aligned} \partial U/\partial q_{it}^{\mathrm{d}} &= \nu p_{it} \\ \partial U/\partial c_t^{\mathrm{d}} &= \nu p_t \qquad \text{all } t \\ \partial U/\partial l_t^{\mathrm{s}} &= -\nu w_t \end{aligned} \tag{11.17}$$

$$\partial U/\partial \nu = \sum_{i=1}^{4} (\pi_i + \pi_{gi}) + \sum_{t=1}^{T} \left(w_t l_t^{\mathrm{s}} - \sum_{i=1}^{3} p_{it} q_{it}^{\mathrm{d}} - p_t c_t^{\mathrm{d}} \right) = 0$$

where ν is the Lagrange multiplier associated with the budget constraint. (The Lagrange multiplier ν is interpreted as the marginal utility of income.) Conditions (11.17) imply that the marginal rate of substitution between any two commodities in the tth period equals their relative price in that period.

11.3 General Equilibrium Cost–Benefit Rules

To derive cost–benefit rules, we need a monetary welfare change measure that captures the social benefits and costs of *small* changes in a policy instrument, i.e. in the level of production of a state owned firm. To obtain such a measure, totally differentiate the utility function (11.15) and substitute the first order utility maximization conditions (11.17):

$$\begin{aligned} \mathrm{d}W = \mathrm{d}U/\nu &= \sum_{t=1}^{T} \left(\sum_{i=1}^{3} (\partial U/\partial q_{it}^{\mathrm{d}})\, \mathrm{d}q_{it}^{\mathrm{d}} + (\partial U/\partial c_t^{\mathrm{d}})\, \mathrm{d}c_t^{\mathrm{d}} \right. \\ &\qquad \left. + (\partial U/\partial l_t^{\mathrm{s}})\, \mathrm{d}l_t^{\mathrm{s}} \right) \Big/ \nu \\ &= \sum_{t=1}^{T} \left(\sum_{i=1}^{3} p_{it}\, \mathrm{d}q_{it}^{\mathrm{d}} + p_t\, \mathrm{d}c_t^{\mathrm{d}} - w_t\, \mathrm{d}l_t^{\mathrm{s}} \right) \end{aligned} \tag{11.18}$$

It is useful to rewrite (11.18) in terms of policy instruments. Differentiating the household budget constraint (11.16) gives

$$\sum_{t=1}^{T} \left(\sum_{i=1}^{3} p_{it}\, \mathrm{d}q_{it}^{\mathrm{d}} + p_t\, \mathrm{d}c_t^{\mathrm{d}} - w_t\, \mathrm{d}l_t^{\mathrm{s}} \right) = \sum_{i=1}^{4} (\mathrm{d}\pi_i + \mathrm{d}\pi_{gi}) + \sum_{t=1}^{T} l_t^{\mathrm{s}}\, \mathrm{d}w_t \tag{11.16a}$$

Observe that all prices except the wage w_t are fixed at world market determined levels, i.e. $\mathrm{d}p_{1t} = \ldots = \mathrm{d}p_{3t} = \mathrm{d}p_t = 0$. Substituting (11.16a) into (11.18) yields

$$\mathrm{d}W = \sum_{i=1}^{4} (\mathrm{d}\pi_i + \mathrm{d}\pi_{gi}) + \sum_{t=1}^{T} l_t^s \,\mathrm{d}w_t \tag{11.19}$$

From the envelope theorem, which was derived in (6.36), it follows that the change in profits of a profit maximizing (i.e. private sector) firm in (11.19) is given by

$$\mathrm{d}\pi_i = \mathrm{d}\bar{\pi}_i - \sum_{t=1}^{T} l_{it}\,\mathrm{d}w_t = -\sum_{t=1}^{T} l_{it}\,\mathrm{d}w_t \tag{11.20}$$

Recall that $\mathrm{d}p_{1t} = \ldots = \mathrm{d}p_{3t} = \mathrm{d}p_t = 0$ owing to the small open economy assumption. $\mathrm{d}\bar{\pi}_i$ denotes the change (equal to zero) in the sum of the discounted profits calculated at *constant* prices and wages. To see that $\mathrm{d}\bar{\pi}_i = 0$ we differentiate (11.1) assuming that all prices and wages are constant:

$$\mathrm{d}\bar{\pi}_1 = \sum_{t=1}^{T} \left(p_{1t}\frac{\partial f_{1t}}{\partial c_{1t}} - p_t\right)\mathrm{d}c_{1t} + \left(p_{1t}\frac{\partial f_{1t}}{\partial l_{1t}} - w_t\right)\mathrm{d}l_{1t} \tag{11.20a}$$

From conditions (11.2) for profit maximization, it is seen that both expressions within ( ) are equal to zero: the value of the marginal product of each factor in the tth period is equal to its price in that period. Hence $\mathrm{d}\bar{\pi}_1 = 0$. Of course, this holds true for any profit maximizing firm, i.e. $\mathrm{d}\bar{\pi}_1 = \mathrm{d}\bar{\pi}_2 = \mathrm{d}\bar{\pi}_3 = \mathrm{d}\bar{\pi}_4 = 0$. On the other hand, by assumption, state owned firms need not be profit maximizers. Hence $\mathrm{d}\bar{\pi}_{gi} \gtreqless 0$ for such firms.

Using this information, the welfare expression (11.19) is reduced to:

$$\mathrm{d}W = \sum_{i=1}^{4} \mathrm{d}\bar{\pi}_{gi} - \sum_{t=1}^{T}\left(\sum_{i=1}^{4} (l_{git} + l_{it}) - l_t^s\right)\mathrm{d}w_t \tag{11.21}$$

$$= \sum_{i=1}^{4} \mathrm{d}\bar{\pi}_{gi}$$

The second term in the first equality in (11.21) vanishes since aggregate demand for labour is equal to its supply when the wage adjusts to achieve full employment in each period. For similar reasons, (11.21) holds (for small changes) even if the exchange rate was flexible (instead of fixed as is assumed in this chapter). We need then add only the balance of trade, which equals zero in each period under flexible exchange rates in the present model, multiplied by the change in the exchange rate. Under a fixed exchange rate regime, the sum over all periods of the trade balance in each period equals zero. This fact, which explains that the trade balance does not appear in (11.21), can be shown by substituting for the profit terms in the household's budget constraint (11.16), and using the fact that the labour market clears in each period. [See also (11.45).] Also note that *fixed* costs – if such costs were explicitly introduced – do not appear in a welfare (or profit) *change* measure like (11.21).

Equation (11.21) gives a monetary measure of the welfare change following from a small change in the level of production or extraction of a certain good or resource (assuming that there exists an equilibrium in both the initial situation and the final situation). This measure can be used provided that the changes in each of the variables concerned are sufficiently small for the marginal utility of income (v) to be treated as constant throughout the movement. See the discussion in section 11.4.3, where we also generalize the welfare measure (11.21) to a multihousehold economy. To interpret the welfare measure, it is convenient to examine various cases more closely.

11.3.1 *Ordinary Produced Goods* (Pulp)

Assume that the state owned firm producing pulp marginally increases its level of production. Using the government budget constraint (11.14), the change in profits – calculated at constant prices and wages – of this firm can be shown to be

$$\begin{aligned} d\bar{\pi}_{g1} &= \sum_{t=1}^{T}\left[\left(p_{1t}\frac{\partial f_{g1t}}{\partial c_{g1t}}-p_t\right)dc_{g1t}+\left(p_{1t}\frac{\partial f_{g1t}}{\partial l_{g1t}}-w_t\right)dl_{g1t}\right] \\ &= \sum_{t=1}^{T}\left[p_{1t}-MC_{1t}(q_{g1t})\right]dq_{g1t} \qquad (11.22) \end{aligned}$$

where MC_{1t} denotes the firm's marginal cost in period t as a function of its level of production q_{g1t} (see section 6.7, and Henderson and Quandt, 1980, chapter 4 for a derivation of the cost function for a firm using two inputs).

From (11.22) and figure 11.2 it should be clear that it is possible to increase the firm's profit in the tth period if the value of the marginal product of a factor (timber or labour) differs from its price in that period [see also (11.20a)]. In other words, to maximize its profits the state owned firm, like private sector firms, should in each period select a level of production such that the price p_{1t} of its output equals the marginal cost MC_{1t} of production in that period. This

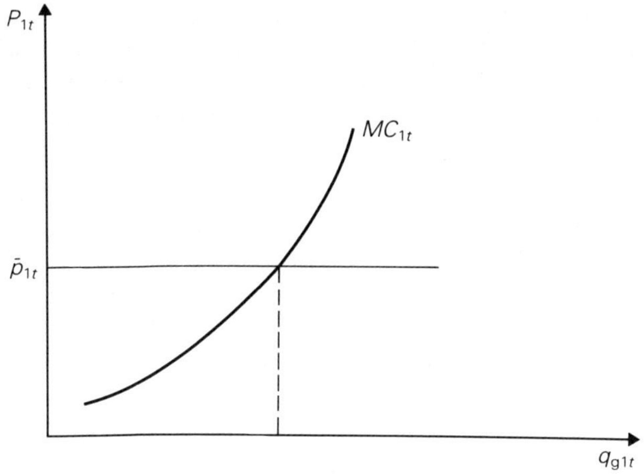

FIGURE 11.2 Increased production of an ordinary produced good increases profit as long as the exogenously fixed world market price $\bar{p}_{1t}$ of the good exceeds the marginal cost MC_{1t} of producing the good

argument is illustrated in figure 11.2. Now, since $dW = d\bar{\pi}_{g1}$ in the welfare expression (11.21), increased production in any period t by the state owned firm will raise welfare as long as $p_{1t} > MC_{1t}$. This is the general equilibrium rule obtained by Boadway (1975, 1978), Boiteux (1956), Harberger (1971), Lesourne (1975) and Starrett (1979) among others.

11.3.2 *Non-renewable Resources*

The simple rule for project evaluations derived in the previous section cannot be used to evaluate (small) changes in the level of extraction of a non-renewable resource. Increased extraction in say the current period requires that extraction in some later period t is reduced (provided, of course, that the exhaustion constraint is binding). Hence, the loss in future marginal profits must be accounted for. Using (11.14) and (11.21) the change in monetary welfare can be shown to be

$$\begin{aligned} dW = d\bar{\pi}_{g2} &= [(p_{21}f'_{g21} - w_1) - (p_{2t}f'_{g2t} - w_t)(f'_{g21}/f'_{g2t})]\ dl_{g21} \\ &= \{[p_{21} - MC_{21}(q_{g21})] - [p_{2t} - MC_{2t}(q_{g2t})]\}\ dq_{g21} \qquad (11.23) \end{aligned}$$

i.e. the change in welfare equals the sum of the increase in period 1 profits and the *loss* in period t profits. Welfare increases (decreases) if the first term exceeds (falls short) of the second term in (11.23).

A geometric description of the two period case is contained in figure 11.3. Owing to the constraint that the firm can extract no more than the initial stock $\bar{x}_{g0}$, the condition $p = MC$ is impossible to fulfil simultaneously in all (both) periods. The best the firm can

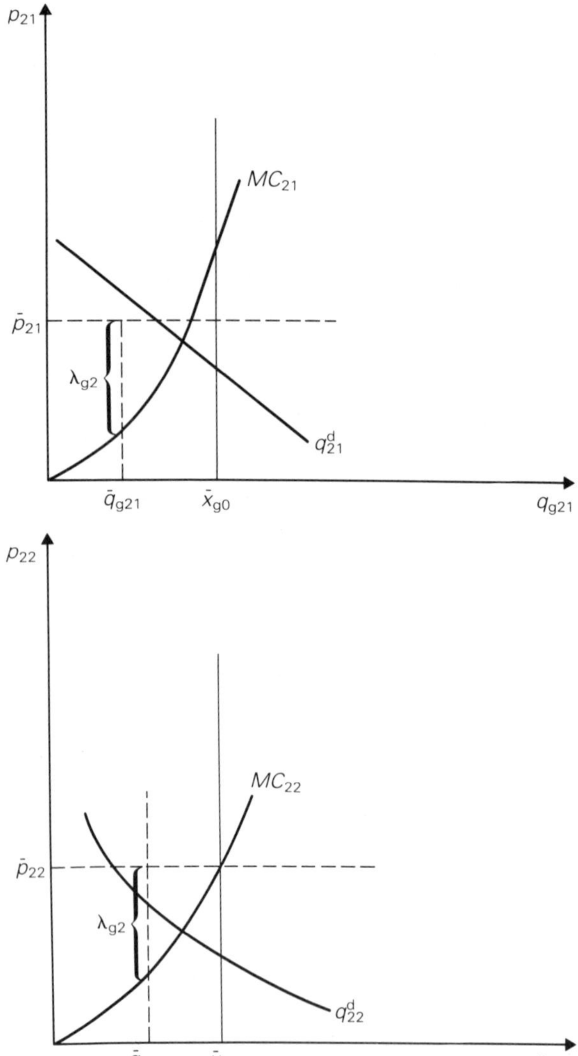

FIGURE 11.3 Optimal levels of extraction of a non-renewable resource in the two period case. The firm can extract ($\bar{q}_{g21} + \bar{q}_{g22}$) units, i.e. its initial stock $\bar{x}_{g0}$ of the resource

do is to allocate extraction between periods in such a way that the positive difference between the prevailing world market determined output price ($\bar{p}_{21}$ and $\bar{p}_{22}$ in the figure) and the marginal extraction cost is the same in all (both) periods.

In effect, this is the condition for profit maximization stated in (11.4). Thus if the state owned firm maximizes its profits and hence also society's welfare, it must hold that $p_{2t} - MC_{2t} = \lambda_{g2} =$ constant for all t, where the constant λ_{g0} may be interpreted as the Lagrange multiplier associated with the constraint that no more than the initial stock $\bar{x}_{g0}$ is extracted. Recall (11.4). In figure 11.3, the firm extracts $\bar{q}_{g21}$ units in period 1 and $\bar{q}_{g22}$ units in period 2 (with $\bar{q}_{g21} + \bar{q}_{g22} = \bar{x}_{g0}$). If the marginal cost of extraction is constant and equal in both periods, however, the (profit maximizing) firm extracts the whole stock in the second period since $\bar{p}_{22} > \bar{p}_{21}$. Finally, note that domestic consumption of the resource is given by the intersection of a price line drawn at the going world market determined price ($\bar{p}_{21}$ and $\bar{p}_{22}$) and the domestic demand curve (q_{21}^{d} and q_{22}^{d}). In figure 11.3, the country is a net importer in period 1 and a net exporter in period 2. In constructing figure 11.3 the level of domestic private sector extraction has for simplicity been set equal to zero. An alternative possibility is to interpret the demand curves as domestic demand less domestic private sector extraction at the going price.

11.3.3 *Renewable Resources*

In order to keep the stock of a renewable resource at any predetermined level, a firm can harvest no more than the yearly growth of the resource. Suppose, however, that the state owned firm in period 1 harvests one unit over and above the natural growth. To keep the stock at this new lower level it has to reduce the harvest in every future period by F_{g}' units, where F_{g}' denotes the marginal growth of the stock in period t. The argument is illustrated in figure 11.4. Using (11.14) and (11.21), the accompanying change in welfare can be shown to be

$$\begin{aligned} \text{d}W = \text{d}\bar{\pi}_{\text{g}3} &= (p_{31} f_{\text{g}3}' - w_1)\left(1 - F_{\text{g}}' \sum_{t=2}^{T} (1+r)^{-t+1}\right) \text{d}l_{\text{g}31} \\ &= [p_{31} - MC_{31}(q_{\text{g}31})]\,(1 - F_{\text{g}}'/r)\,\text{d}q_{\text{g}31} \qquad (11.24) \end{aligned}$$

where both the current price and the wage are assumed to be constant over time [i.e. $p_{3t} = p_{31}(1+r)^{-t+1}$ etc.], and the time horizon T is assumed to approach infinity since the series $\Sigma(1+r)^{-t+1}$ is convergent with sum to infinity $1/r$.

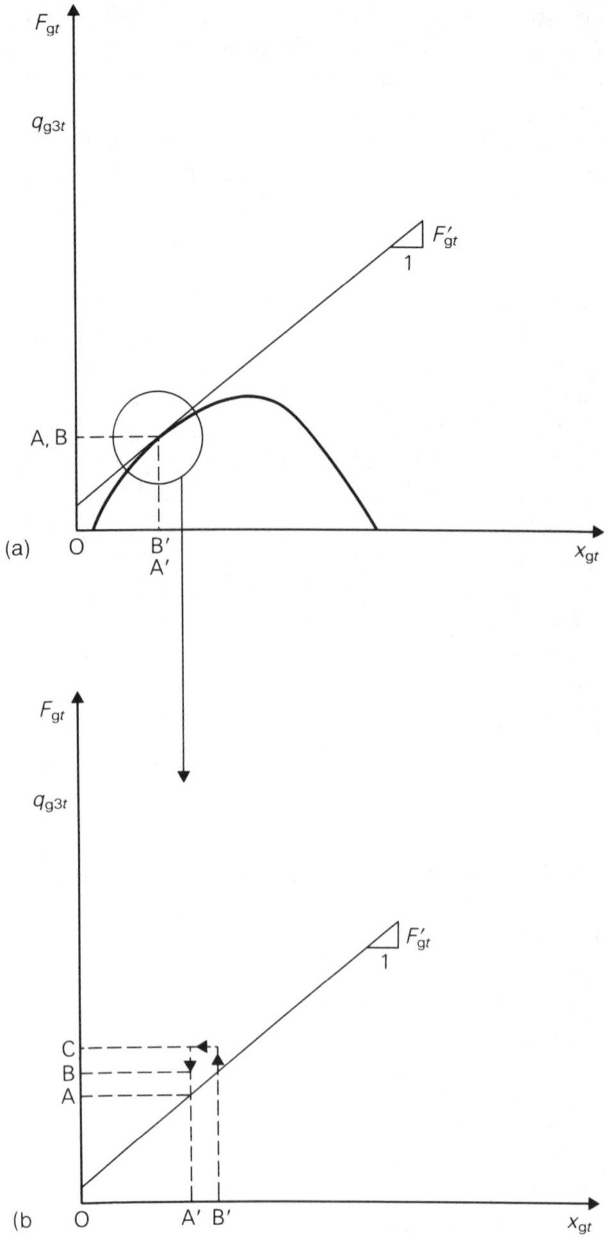

FIGURE 11.4 (a) Illustration of a changed harvest policy. (b) Magnified picture of the neighbourhood of the equilibrium in (a). The initial stock OB′ permits a steady state harvest of OB units but suddenly the firm harvests OC units in a period. This reduces the stock to OA′ units (equal to the initial stock minus BC units) which permits a steady state harvest of OA units

Two conflicting forces are in operation in the welfare measure (11.24). Firstly, consumption of the resource increases in period 1. This is given by the difference between the consumer's price of the output (the marginal willingness to pay) and the marginal cost of harvesting the resource (the marginal disutility of work effort). Provided the natural growth of the resource is a binding constraint on the level of harvest and consumption, there will be a positive difference between the price p_{31} and the marginal harvesting cost MC_{31}. [See conditions (11.7), where the difference equals λ_{31}.] Secondly, to maintain the stock at the new lower level (i.e. to avoid further reductions) consumption must be reduced by F_g' units in each and every future period. The accompanying welfare loss is evaluated in the same way as for increased consumption, i.e. by the difference between the goods price and the marginal harvesting cost.

From (11.24) it should be obvious that welfare is maximized if the (state owned) firm keeps the stock at such a level that $F_g' = r$. This is also the condition for profit maximization derived in (11.7a) and (11.8a). The reader may want to confirm that we would arrive at the same conclusion if the firm harvests an extra unit in period 1 and then reduces its harvest by $(1 + F_g')$ units in, say, period 2 to restore the initial stock of the resource. This policy enables the firm to harvest the initial amount in all future periods ($t = 3, \ldots, T$) and is equivalent from a welfare point of view to the policy considered earlier. Using the welfare measure (11.24) it should also be an easy task for the reader to show that both a free access fishery and the static optimal policy $p_3 = MC_3$ derived in section 3.3 are inferior from a welfare point of view to the optimal policy ($F_g' = r$) derived here. (If the marginal cost of extraction becomes 'very high' at a low level of extraction, however, the natural growth of the resource may constitute no constraint on the yearly harvest. If so, it is optimal to harvest such an amount that $p_{3t} = MC_{3t}$ in all periods, which is also the solution derived in section 3.3.2.)

11.3.4 *Timber Production*

To obtain rules for the management of (national) forests it is convenient to shorten a single (the first) rotation from T_g years to $T_g - 1$ years. By cutting this year a stand that ought to be cut next year, the timber firm pays an opportunity cost in the form of (value) growth forgone. On the other hand, all future net receipts will be realized one year earlier than in the initial cutting programme. Assuming that the current values of the timber price and the wage are constant over time, the resulting change in welfare is given by

$$dW = d\bar{\pi}_{g4} = (p_2 c_{g2} - w_2 l_{g42})(1+r)(1+K_g)^{-1} - (p_2 c_{g2} - w_2 l_{g42}) + (p_2 c_{g2} - w_2 l_{g42})\, r/[(1+r)^{T_g} - 1] \quad (11.25)$$

The first term in this expression is the present value of the profits obtained if the stand is cut in period 1. The second term represents the present value of profits obtained if the stand was cut in the second period instead. The difference between these two terms is due to the fact that trees grow at a rate K_g from period 1 to period 2 and to the discount factor (assuming that the change in the cutting behaviour is small and the periods are short so that the marginal growth rate K_g of trees can be treated as a constant). The sum of these two terms is positive (negative) if the growth rate K_g falls short of (exceeds) the interest rate r. This is shown by the AA locus in figure 11.5. The third term, which is depicted by the BB locus in

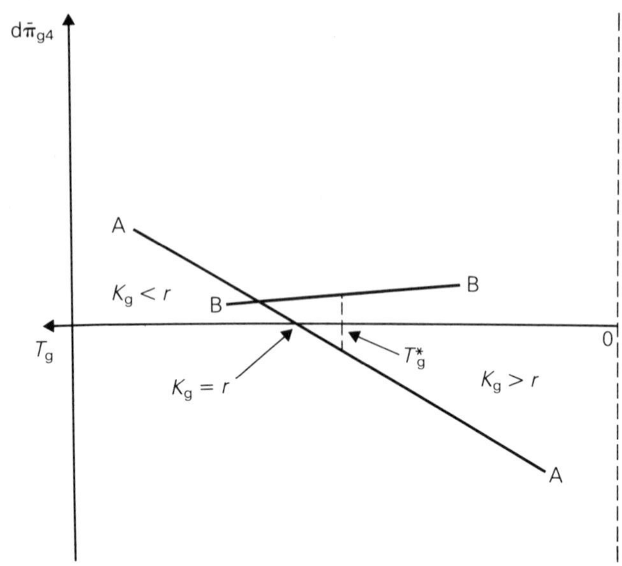

FIGURE 11.5 The AA locus illustrates how the sum of the first two terms in (11.25) varies with the growth rate K_g of trees, provided $dK_g/dT_g < 0$. The BB locus depicts the fact that the third term in (11.25) will be positive if the first rotation period is shortened (since all future net revenues will be realized earlier than before). The sum of these changes is equal to zero ($d\bar{\pi}_{g4} = 0$) if the rotation period is T_g^* years (so that $K_g = K_g^*$); and $-d\bar{\pi}_{g4} \gtrless 0$ as $T_g \gtrless T_g^*$

figure 11.5, captures the increase in the present value of profits when every future cutting takes place one year earlier than the initial programme. It is impossible to further increase welfare (and profits) by changing the cutting programme when the sum of the three terms in (11.25) is equal to zero, i.e. $\mathrm{d}W = \mathrm{d}\bar{\pi}_{g4} = 0$. A comparison between the welfare expression (11.25) and the Faustmann criterion (11.13) reveals that cutting the stand in period 1 instead of in period 2 increases (decreases) welfare if $K_g < K_g^* \, (K_g > K_g^*)$. In turn, this means that welfare can be increased by reallocating cutting decisions between periods as long as the actual growth rate of trees, K_g, differs from the Faustmann growth rate K_g^*. In other words, the profit maximizing Faustmann rotation is also the uniform rotation that maximizes the society's welfare (see also figure 11.5).

In sum, we have the following general equilibrium cost–benefit rule to be used by public sector firms:

> **Proposition 11.1**
> Given the model specified in section 11.2, a public firm that aims at maximizing welfare should use market prices as shadow prices. This holds true regardless of whether the firm produces an ordinary produced good (e.g. pulp), extracts a non-renewable resource or harvests a renewable resource like a forest.

We have already shown that $\mathrm{d}W = 0$ is a necessary condition for welfare to be maximized. Throughout the chapter it is assumed that the second order condition $\mathrm{d}^2W < 0$ for a maximum is fulfilled and that this maximum is unique. For a discussion of the problems that arise in the case of a multiplicity of maxima, the reader is referred to Dasgupta (1982).

It is important to stress that proposition 11.1 does not imply that a natural resource should be managed according to the traditional general equilibrium cost–benefit rule found in textbooks on applied welfare economics. This rule states that a firm should expand its level of production as long as the cost of producing an additional unit falls short of the output price. The incorrectness of this rule, when applied to a renewable or non-renewable resource, should be obvious from the cases examined earlier.

In this section we have derived cost–benefit rules to be used in a small open economy, i.e. an economy that treats all prices but the nominal wage as world market determined. However, the same cost–benefit rules could be derived for a closed economy, i.e. an economy where all prices adjust so as to achieve equality between domestic supplies and demands. This claim will not be proved here, but should be fairly obvious from the results derived in chapter 10 on the properties of a Pareto sustaining price system.

11.4 Multiple Use of Forest Lands

National forests are often managed in ways which depart from the Faustmann rotation criterion. This is true, at least in part, because national forest lands are managed also for non-timber purposes like recreation, wilderness and wildlife. The market is inoperative for many of these multiple uses of forest land. This provides justification for national forest managers to depart from *market* efficiency criteria in their timber management.

In previous chapters, multiple use has been modelled by including the stock of a resource as a separate argument in the utility function. Matters become a bit more difficult when one turns to a real world resource like forest land. There are many alternative uses of forest lands; wood production and harvest are more or less compatible with some of these uses and completely incompatible with others. The most obvious example of incompatible uses is that modern wood production destroys wilderness, and wilderness is valued by hikers and campers among others.[4]

11.4.1 *A Virgin Forest*

Suppose the public timber producer considers cutting down a virgin forest. In addition to the purely economic consequences we must now include the loss of 'non-timber outputs' for the current generation and some, possibly all, future generations. To harvest the forest and use the land for efficient wood production is socially profitable if the discounted sum of the current profit $\bar{\pi}^c_{g41}$ from harvesting the forest plus the current profits $\bar{\pi}^c_{g4T^*_g}$ from all future (Faustmann) rotations, if positive, exceeds the total willingness to pay Y for having the virgin forest saved:[5]

$$\mathrm{d}W_M = \bar{\pi}^c_{g41} + \bar{\pi}^c_{g4T^*_g}/[(1+r)^{T^*_g} - 1] - Y > 0 \tag{11.26}$$

[4] See also the literature on option value e.g. Fisher and Hanemann (1983), Hanemann (1984) and Weisbrod (1964).

[5] More generally this approach can be used to rank the profitability of any alternative uses of land (recreation, agriculture, timber production etc.). See chapter 1 for relevant criteria for ranking different investment opportunities. However, see also the discussion in section 11.4.3.

where a superscript c is used to denote profits in current terms. If management and harvesting of the standing timber is profitable, while continued timber management is inefficient ($\bar{\pi}^{c}_{g4T^*_g} < 0$), the second term must be deleted from (11.26). This is known as the timber mining case. It should also be stressed that $dW_M < 0$ in (11.26) does not necessarily mean that the virgin forest should be saved to the end of the planning horizon. Rather it says that it is socially unprofitable to cut the forest in the first period. *A priori* one cannot exclude the possibility that it will be socially profitable to cut the virgin forest in some later period. This could be the case if the subjective value or attractiveness of virgin forests decreases over time, e.g. if future generations impute no value to wilderness. Furthermore, note that only the virgin forest itself is valued in (11.26). In section 11.4.2 we will take up the case where the age structure of a forest has an influence on social welfare.

The nature of the variable Y in (11.26) is best understood if the land containing the virgin forest $\bar{C}_0$ is included as a separate argument (for each period t, $t = 1, \ldots, T$) in society's welfare function:

$$U_M = U_M(q^d_{11}, \ldots, l^s_T, \bar{C}_{01}, \ldots, \bar{C}_{0T}) \qquad (11.27)$$

In (11.27) the land containing the virgin forest is dated to highlight the fact that its (non-market) value may vary over time.

The total value of having the virgin forest saved to the end of the planning horizon is obtained by calculating the sum of the value of the forest in each period:

$$\sum_{t=1}^{T} \int_0^{\bar{C}_{0t}} (1/v)\, U'_M\, (q^d_{11}, \ldots, l^s_T, \bar{C}_{01}, \ldots, C_{0t}, \ldots, \bar{C}_{0T})\, dC_{0t} \qquad (11.28)$$

where U'_M denotes the partial derivative of the utility function with respect to C_{0t}. Loosely speaking, one might interpret the integral in (11.28) as corresponding to the area between zero and $\bar{C}_{0t}$ under a properly defined 'demand' curve for virgin forests. [In a multihousehold economy the individual 'demand' curves must be added (vertically) to obtain the relevant area.] This area gives the total willingness to pay Y_t for having the virgin forest preserved in the tth period. Summing over all periods, we arrive at the total willingness to pay Y used in the welfare measure (11.26). The reader should observe, however, that from a theoretical point of view there are some severe problems in defining discrete welfare change measures of this kind.[6] We address these problems in section 11.4.3.

[6] In fact, the money measure (11.26) is proportional to the change in utility if, and only if, $\partial v/\partial C_{0t} = \partial v/\partial \bar{\pi}_t = 0$ for all t.

11.4.2 *The Age Structure of Trees Affects Utility*

Another kind of 'external effect' is present if the age of trees affects societal welfare in one way or the other. For example, the attractiveness of the environment may depend on the age distribution of forests. We know, for example, that campers and hikers enjoy the tallest groves of trees (see Brown, 1977). One possible way of modelling this case is by distinguishing between different age classes of trees in the utility function.

$$U_{\mathrm{M}} = U_{\mathrm{M}}(q_{11}^{\mathrm{d}}, \ldots, l_T^{\mathrm{s}}, C^1, \ldots, C^T) \tag{11.29}$$

where $C^t = (\tilde{c}_0^t, \ldots, \tilde{c}_\Upsilon^t)$ all t, and $\tilde{c}_i^t$ represents the number of trees aged i years in the tth period ($t = 1, \ldots, T$). If there are X_{i-1}^{t-1} trees of age $i-1$ years in period $t-1$ and c_{i-1}^{t-1} of these trees are cut down, then there remain $\tilde{c}_i^t = X_{i-1}^{t-1} - c_{i-1}^{t-1}$ trees aged i years in period t. The vector C^t is contained in the utility function to represent the 'subjective' valuation in any period t of trees (or land containing trees) of different ages. Trees (or forest land containing trees) older than Υ years are not valued at all, although Υ may be a very large number. It can be reasonably assumed that the marginal utility $\partial U_{\mathrm{M}}/\partial \tilde{c}_i^t$ of saving a tree aged $i-1$ years for another year – instead of cutting it down – is positive but diminishing (for $i \leqslant \Upsilon$).

Next, if we proceed in the same way as in section 11.3, the relevant welfare criterion for evaluating a small shortening of the uniform rotation period T_{g} for the national timber producer can be shown to be[7]

$$\mathrm{d}W_{\mathrm{M}} = \mathrm{d}\bar{\pi}_{\mathrm{g}4} - \sum_{t=2}^{T} (\partial U_{\mathrm{M}}/\partial \tilde{c}_{T_{\mathrm{g}}}^t v)\, \mathrm{d}c_{\mathrm{g}1}^{t-1} \tag{11.30}$$

In (11.30) *all* rotations are shortened. Using the results derived in section 11.3, it should be straightforward to show that a marginal shortening of all rotations, like a shortening of a single rotation, leaves $\bar{\pi}_{\mathrm{g}4}$ unchanged if the firm maximizes profits in the initial situation. Hence, the first term vanishes if the firm harvests according to the Faustmann principle. The second term gives the loss in utility

[7] Equation (11.30) is obtained by adding the sum of external effects

$$\left[- \sum_{t=2}^{T} (\partial U_{\mathrm{M}}/\partial \tilde{c}_{T_{\mathrm{g}}}^t v)\, \mathrm{d}c_{\mathrm{g}1}^{t-1} \right]$$

to (11.18), (11.19) and (11.21) and assuming that the levels of public sector production other than timber harvesting are kept constant ($\mathrm{d}\bar{\pi}_{\mathrm{g}1} = \mathrm{d}\bar{\pi}_{\mathrm{g}2} = \mathrm{d}\bar{\pi}_{\mathrm{g}3} = 0$).

due to the shortening of the rotation periods when a higher utility is imputed to older forests than to younger ones. Or, to put it the other way round, it represents the society's (or the sum of each individual's) marginal willingness to pay for having a stand saved for an additional year. From this it is clear that the Faustmann rotation T_g^* is shorter than the socially optimal rotation period whenever the utility function takes the form (11.29), and $\Upsilon > T_g^*$. Hence, it is socially profitable to lengthen the rotation period as long as the loss in profits $\bar{\pi}_{g4}$ of a further (small) lengthening falls short of the sum of the positive marginal willingness to pay $(\partial U_M/\partial \bar{c}^t \nu)$ for such a lengthening. From a welfare point of view, the rotation period should be such that a marginal shortening/lengthening of the period leaves the *sum* of the terms in (11.30) unchanged, i.e. $dW_M = 0$, assuming such an interior solution exists.

Note, however, that the measure (11.30) neglects a possibly important effect. A shortening of the rotation period means that any stand in any future rotation will reach a certain age earlier than in the initial programme, e.g. in 2084 instead of in 2085. This 'shift', which may have a positive or negative impact on welfare depending on the assumed properties of the utility function, has not been included in (11.30). Suppose that the shift has a positive impact on welfare. This could be the case if 'consumption' of old trees today is preferred to consumption of such trees tomorrow. Then there is a possibility that this positive effect of a shortened rotation exceeds the second and negative term in (11.30). If so, the socially optimal rotation will in fact be *shorter* than the Faustmann rotation![8]

In this section, we have concentrated on multiple uses or external effects that directly affect utility. The harvest policy may, of course, also have an impact on the production possibilities of other firms in the economy. This can be modelled by including, say, the harvested volume or the age of forests as separate arguments in the production functions. It is left as an exercise to the reader to perform the analysis, if necessary, after having consulted section 3.3 which deals with one particular kind of external effect in production.

In sum, a properly conducted cost-benefit analysis of the effects of cutting down a certain stand or shortening/lengthening the rotation period must recognize the possibility of multiple use. Theoretical exercises of the kind performed here may give an idea of which effects to account for. Needless to say, it is quite another matter to estimate such effects in real life.

[8] A similar result has been obtained by Hartman (1976). See also Kaiser and Norbury (1984).

11.4.3 *The Taking Issue in Forestry*

From the analysis so far it is clear that a profit maximizing (private or public) timber producer cannot always be expected to behave in a socially optimal way if there are forest uses other than timbering. This raises the important question of whether forest owners should be induced ('bribed') or forced through regulations to behave in the 'right way'. Either approach could in principle be used to ensure the socially optimal levels of different outputs from forest lands. However, as is demonstrated by the renewed interest in this so-called *taking* issue in the US, it is far from uninteresting which principle is used.[9] Taking requires compensation according to the US Constitution, while regulation does not. Of course, the government could use other measures (e.g. taxation as in section 3.3) to secure the socially optimal levels of production, but we will concentrate on the taking issue here.

To be able to discuss the taking issue in a meaningful way, we abandon our single household assumption and explicitly consider a multihousehold economy. In a multihousehold economy it is not sufficient that $dW > 0$ for a project to be Pareto improving. This is because some households may gain and some households may lose from the project considered. For the project to lead to Pareto improvements, the gainers must compensate the losers so that nobody is left worse off than before the project. It can be shown that the following conditions are necessary and sufficient to ensure that a project is Pareto improving in a multihousehold economy:[10]

Proposition 11.2
A marginal project can be instituted in terms of strict Pareto improvements iff the aggregate monetary welfare change measure has a positive sign:

$$dW = \sum_{n=1}^{N} dW^n > 0$$

where $dW^n \gtreqless 0$ for any household n $(n = 1, \ldots, N)$.

[9] For an instructive analysis of the taking issue in a risky world see Blume, Rubinfeld and Shapiro (1984).

[10] See e.g. Starrett (1979). However, a rule or a constitution for deriving, from individual orderings of social states, a social ordering consistent with some reasonable conditions cannot be found in general (Ng, 1979, p. 112). In simple terms, this is Arrow's (im)possibility theorem (see Arrow, 1951).

In order to guarantee strict Pareto improvements for the project, gainers must compensate losers so that everyone is made better off:

$$dW^n + \delta^n > 0 \qquad \text{all } n$$

where $\delta^n \gtrless 0$ represents a lump sum payment to (from) the nth household and

$$\sum_{n=1}^{N} \delta^n = 0$$

Regulation means that the government imposes constraints on the firm's maximization problem. For example, a timber producer may not be allowed to cut a stand before it reaches a certain prespecified age (which presumably exceeds the Faustmann rotation period). Alternatively some productive timberland may be reserved for non-timber multiple use. In any case, regulation *per se* means that the owner of the regulated area has to bear the costs, while the benefits, as a rule, accrue to others. Hence, regulation may very well fulfil the first part of the Pareto criterion ($dW > 0$), but fails to fulfil the second part of the criterion ($dW^n + \delta^n > 0$ for all n). However, it is obviously possible to combine regulation with taxation according to the *benefits principle* – that is, tax the gainers of a certain regulation and use the revenue to compensate the losers. If so, regulation can be used to improve welfare according to the Pareto criterion when there are multiple uses of forest lands.

Taking (according to the US Constitution) may seem to fulfil the Pareto criterion since forest land owners are compensated for economic losses caused by land use restrictions. This assumes, however, that taxation is according to the benefits principle so that the gainers of a restriction are those who actually pay the compensation. Recall that the Pareto criterion requires that nobody loses from a change considered. If the compensation is paid by the 'ordinary' taxpayer, who neither gains nor loses from the project as such, taking will not pass the Pareto test in proposition 11.2. This fact seems to have been overlooked in the recent US debate on the taking issue (see Sedjo, 1983). Taking is supposed to be superior to regulation since forest land owners are compensated in the former case but not in the latter. However, taking may (but need not) mean that the burden is simply passed over from one group of households to another, while the benefits accrue to a third group of households. Of course, nothing ensures that such a 'distribution scheme' is Pareto superior to regulation without compensation.

In closing, it should be stressed that the evaluation of discrete or large policy changes is still an unresolved issue in welfare economics. Loosely speaking, if a policy change more than marginally affects the marginal utility of income of any individual, then there exists in general no unique money measure of utility change. This is because the size of the change ΔW^n will depend on the particular path of integration between the terminal situations (see Silberberg, 1972 and Morey, 1984). In order to circumvent this problem, cost–benefit analysts often assume that each household is kept on its initial or preproject level of utility. If the sum of the resulting so-called compensating variations of the individuals is negative,[11] a project is assumed to lead to welfare improvements. However, there exists no simple relationship between this criterion and the ability to compensate losers if changes are discrete. It would take us too far to show this, so we have to refer the interested reader to Boadway (1974), Boadway and Bruce (1984), Just et al. (1982) and Smith and Stephen (1975) for discussions of this so-called Boadway paradox.

In effect, the only useful welfare change measure we can find based on the Pareto principle is one referring to infinitesimal changes, e.g. that in (11.21). Reliance on the Pareto criterion when evaluating large changes, like the harvest of a virgin forest in (11.26), cannot be justified unless we make very restrictive assumptions regarding the properties of the utility functions. Some distributive weights must be chosen to apply to the gains and losses of each individual. This choice must be made on non-economic grounds.

11.5 Socially Optimal Cutting Rules in Unemployment Situations

In section 11.4 we considered one reason (multiple use) for managers of national forests to depart from the classical Faustmann principle in their forest management. Another important reason for not adopting the Faustmann rule is that the public timber supply policy is often designed to minimize the variation in employment over time for those whose employment is directly based upon national forest timber sales. This seems to be true not only in Western Europe but also in Canada and the United States. (See Bourchier, 1982; Bränn-

[11] The hypothetical income change that keeps an individual on the initial or preproject level of utility is called the compensating variation. We adopt the notational convention that this income change is negative for gainers and positive for losers, implying that welfare can be increased if the sum is negative.

mann, 1984; Howe, 1979, chapter 11; Hyde, 1980, chapter 2 and Jackson, 1980, chapter 5.)

In this section, we examine whether unemployment is justification for temporary deviations from the Faustmann rule. In order to concentrate on the market imbalance issue while setting aside matters of multiple uses of forest land, it is assumed that forest lands are managed exclusively for timber production. Firstly, cost-benefit rules are derived to be used when evaluating the efficiency gains from increased cutting in a situation where the economy is suffering from so-called *classical unemployment*, to use a term familiar from modern macroeconomics (see Cuddington, Johansson and Löfgren, 1984 or Malinvaud, 1977). In this situation, there is unemployment due to excessive real wages while firms are unconstrained in all markets.[12] Then we turn to a situation which is labelled *Keynesian unemployment*. In this case there is unemployment and (timber) producers face a sales constraint on their final output owing to insufficient demand.

11.5.1 *Classical Unemployment*

Assume that the wage in a certain period is fixed at a level causing an excess supply of labour (unemployment) (see figure 11.6). The fact that there is unemployment but no other market imbalances means that firms are unconstrained in all markets. Hence, the maximization problems of firms are those described in section 11.2. On the other hand, the representative household now maximizes its utility function (11.15) subject to the budger constraint (11.16) *plus* the employment constraints

$$l_t^{\mathrm{s}} \leqslant \bar{l}_t = \sum_{i=1}^{4} (l_{it} + l_{git}) = l_t^{\mathrm{d}} \qquad \text{all } t \tag{11.31}$$

where $\bar{l}_t$ is the constraint on the level of employment in the tth period. The constraint is binding if (11.31) holds with equality. The first order condition for utility maximization with respect to l_t^s in (11.17) is replaced by the following equation:

$$\partial U/\partial l_t^s = -\nu w_t + \theta_t \qquad \text{all } t \tag{11.32}$$

where $\theta_t \geqslant 0$ is the Lagrange multiplier associated with the constraint in the labour market in the tth period. Using this new information

[12] In a closed economy classical unemployment refers to a situation where households are rationed in both goods markets and labour markets. See Cuddington Johansson and Löfgren, 1984, chapters 2 and 3, and Marks, 1979.

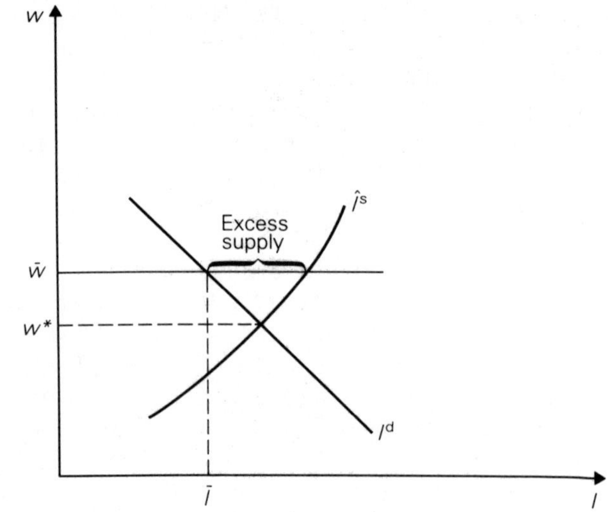

FIGURE 11.6 The labour market when the wage is fixed above the market clearing level. $\hat{l}^s$ denotes notional supply of labour and l^d notional demand for labour. The level of employment is demand determined and equal to $\bar{l}$

and assuming that the employment constraint is binding in all periods, (11.18) is replaced by

$$dW_u = dU/v = \sum_{t=1}^{T}\left(\sum_{i=1}^{3} (\partial U/\partial q_{it}^d)\, dq_{it}^d + (\partial U/\partial c_t^d)\, dc_t^d + (\partial U/\partial l_t^s)\, d\bar{l}_t\right)\bigg/ v$$

$$= \sum_{t=1}^{T}\left(\sum_{i=1}^{3} p_{it}\, dq_{it}^d + p_t\, dc_t^d - w_t\, d\bar{l}_t\right) + \sum_{t=1}^{T} (\theta_t/v)\, d\bar{l}_t \tag{11.18a}$$

Substituting (11.16a) into (11.18a),[13] assuming that the levels of public sector production other than timber harvesting are fixed ($d\bar{\pi}_{g1} = d\bar{\pi}_{g2} = d\bar{\pi}_{g3} = 0$), we obtain

$$dW_u = d\bar{\pi}_{g4} + \sum_{t=1}^{T} (\theta_t/v)\, d\bar{l}_t \tag{11.33}$$

[13] Note that the last term in (11.16a) vanishes since we now assume that $dw_t = 0$ for all t. Similarly, $d\pi_i = 0$ when $dw_t = 0$ for all t, as seen in (11.20).

Equation (11.33) provides a straightforward rule for evaluating policy induced changes in timber production under classical unemployment. First, evaluate timber output and labour inputs at prevailing domestic market prices and assess the change in the public timber firm's profitability. Note that private profit maximizing timber producers do not change their cutting decisions since all prices are fixed. Second, determine the change in the level of employment of the public timber producer in each period ($d\bar{l}_t = dl_{g4t}$) and evaluate this change at the prevailing wage less any adjustment reflecting household disutility of increased work effort. In other words, the Lagrange expression θ_t/ν may be interpreted as the amount of compensation that can be taken from the household while leaving it just as well off as it was before the change in employment.

The welfare measure in (11.33) differs from the general equilibrium measure (11.25) in that a term reflecting the change in employment is now added. If the wage is allowed to adjust so as to achieve full employment, the second term in (11.33) would be equal to zero. The reason is that the employment constraints (11.31) do not 'bite' in this case. On the other hand, if unemployment is excessive or 'chronic' it seems reasonable to set θ_t/ν equal to the wage rate w_t.[14] In effect, this means that household disutility of increased work effort is zero ($\partial U/\partial l_t^s = 0$). This is a commonly used assumption in empirical project evaluation owing to lack of data on the perceived disutility of additional employment. In what follows, we introduce the more general assumption that the social cost (in current terms) of employing an otherwise unemployed labourer is non-negative but constant ($\alpha \geqslant 0$) in all periods:

$$\alpha = -(1+r)^t(\partial U/\partial l_t^s)/\nu = w_t - (1+r)^t\theta_t/\nu \geqslant 0$$

This assumption is now employed in a more detailed analysis of the benefits and costs of increased cutting 'today' (followed by decreased cutting 'tomorrow'). The reason for increasing cutting and hence employment today may be a political judgement that today's level of unemployment is excessive relative to tomorrow's expected level. Two 'extreme' cases – persistent unemployment and temporary unemployment – will be examined. Most real world situations probably fall somewhere between these extreme cases. However, the two cases considered in the following should be useful in crystallizing the

[14] If $(\theta_t/\nu) = w_t$ for all t, labour costs can be eliminated from (11.33), i.e.

$$dW_u = p^c\, dc_g/[1-(1+r)^{-T_g}]$$

assuming that timber prices p^c in current terms remain constant over time.

main effects which have to be taken into account in empirical investigations. Consider first the persistent unemployment case.

Persistent unemployment. From section 11.3 it should be clear that increased cutting this year by the public timber firm followed by decreased cutting next year leaves the present value of the firm's profits unchanged, provided all prices and wages are expected to be constant. This will hold if it is assumed that the firm returns to a Faustmann cutting programme in the third year (and then follows this profit maximizing programme forever). Then, the first term ($d\bar{\pi}_{g4}$) in the welfare change measure (11.33) vanishes, and the expression can be rewritten as

$$\begin{aligned} dW_u &= \sum_{t=1}^{T} (\theta_t/v)\, d\bar{l}_t = \sum_{t=1}^{T} (w-\alpha)(1+r)^{-t} dl_{g4t} \\ &= (w-\alpha)\,[dl_{g41} - (1+K_g^*)\, dl_{g41}(1+r)^{-1}] \\ &\quad + (w-\alpha)(1+K_g^*)\,[dl_{g41}(1+r)^{-T_g^*} - dl_{g41}(1+r)^{-T_g^*-1} \\ &\quad + \ldots] \end{aligned} \tag{11.34}$$

where it has been assumed that current timber prices and wages remain constant over time, T_g^* is the uniform profit maximizing (Faustmann) rotation period, and K_g^* is the marginal growth rate of trees when they have an age of T_g^* years.

To interpret this expression, consider the last equality. By assumption, the firm increases cutting in the first period. The present value to society of the resulting increase in employment dl is equal to $(w-\alpha)\, dl$, ignoring subscripts for the moment. Then the firm reduces employment in the second period since a stand cannot be cut twice. The present value of the resulting welfare loss is equal to $-(w-\alpha)(1+K^*)\, dl(1+r)^{-1}$. Recall that employment is assumed to be proportional to the volume harvested and trees grow at a rate K^*. This explains that the second period loss in employment is $-(1+K^*)\, dl$. Moreover, since the first rotation is shortened, all future rotations occur earlier than before. This causes a sort of positive discounting effect on the value of future employment. This effects is captured by the last term in (11.34).

The sum of the three considered terms can be shown to be

$$dW_u = -(w-\alpha)\, dl_{g41}\,\{K_g^* - r - r/[(1+r)^{T_g^*-1} - 1]\} \tag{11.34a}$$

The terms within { } in this expression sum to zero. This is because $K_g^* = r + r/[(1+r)^{-T_g^*-1} - 1]$ for a profit maximizing producer [see

(11.13a)]. Hence monetary welfare is left constant by the considered marginal change in the profit maximizing cutting programme. (Under a flexible exchange rate regime, however, there would be an additional term in (11.34a) reflecting the change in aggregate employment caused by an appreciation or a depreciation of the currency, provided, of course, that a change in a policy instrument has an impact on the exchange rate. This also holds true for the measures to be derived later.)

Equation (11.34a) indicates that *a high level of unemployment today is not a sufficient condition to justify marginal departures from the Faustmann solution* to the forest management problem. This result is surprising, given that it is often argued that public timber managers should depart from this rule as they have a responsibility for those whose employment is based on national forest timber sales (see Jackson, 1980, pp. 105–7 for details). Such arguments are probably based on cost–benefit rules of the kind found in textbooks, saying that it is socially profitable to expand production beyond the profit maximizing level if there is unemployment in the economy. Unfortunately, this (single period) rule cannot be applied to a natural resource like a forest, which requires an explicit intertemporal treatment; the value of increased production (cutting) and employment 'today' must be weighed against losses of production and employment in future periods.

It should be stressed that throughout this chapter it is assumed that cutting costs are proportional to the volume harvested. This assumption is strategic in deriving the cost–benefit result (11.34a). If cutting costs are not proportional to the harvest level, welfare may very well increase or decrease when the rotation is marginally shortened (starting as before from the profit maximizing rotation period). Given the considered policy change, the second term in the general welfare change measure (11.33) need not be equal to zero for cutting technologies other than that with constant returns to scale.[15] It remains, however, to derive detailed cost–benefit rules for such non-proportional cutting technologies.

Temporary departures from full employment. Now suppose the economy is suffering from unemployment today but is expected to attain full employment next year and to remain in a state of full employment in all future periods. In other words, the nominal wage

[15] See section 6.7 for definitions of technologies with constant, decreasing and increasing returns to scale.

adjusts to achieve full employment in all periods except the present one. If the public timber producer decides to increase cutting today, one would *a priori* expect this to cause wages to *fall* next year. The public sector firm reduces its level of cutting and hence demand for labour next year (since it has already cut the trees that ought to be cut next year). In turn, the reduced labour demand causes a drop in the equilibrium wage rate. The lower nominal wage induces private profit maximizing timber firms to postpone some cutting decisions from this year to next[16] (if they have perfect information or at least rational expectations, as is in fact assumed in deriving the Faustmann rule). Neglecting all future policy induced wage changes (i.e. assuming that the market clearing wage in current terms is $w_3^c = w_4^c = w_T^c = w$), the monetary welfare change can be written

$$\begin{aligned} dW_u &= (w - \alpha)(dl_{g41} + dl_{41}) + \sum_{i=1}^{4} (d\pi_i + d\pi_{gi}) + dw_2 l_2^s \\ &= (w - \alpha)(dl_{g41} + dl_{41}) \end{aligned} \tag{11.35}$$

where $d\pi_i = - dw_2 l_{i2}$ and $d\pi_{gi} = - dw_2 l_{gi2}$ for a profit maximizing firm from (11.20). Recall that all goods prices are fixed owing to the small open economy assumption and that $dw_t = 0$ for $t \neq 2$. Furthermore

$$l_2^s = \sum_{i=1}^{4} (l_{i2} + l_{gi2}) = l_2^d$$

when there is full employment, which explains that the wage change terms sum to zero in (11.35).

In trying to ascertain the sign of the welfare change, it must be recognized that two counteracting forces are in operation. First, the public timber producer increases its level of employment l_{g41} in the period the economy suffers from unemployment. This increases welfare because the difference $(w - \theta)$ between the nominal wage and the marginal disutility of effort is positive owing to unemployment. Second, the increase in public timber output 'crowds out' private timber output in the year with unemployment (as explained earlier). The accompanying drop in private sector employment l_{41} reduces monetary welfare.[17] Although the latter effect is presumably negli-

[16] This increase in demand for labour may partially offset the decrease in the equilibrium wage.

[17] Also private sector firms harvesting other kinds of natural resources may be affected by the drop in the wage rate. This effect has not been included in (11.35).

gible for small changes in the policy instrument, it hints at an important possibility. The behaviour of a large public sector timber producer may have a significant influence on the price expectations of private sector producers. If the public timber producer increases its harvest in periods with high unemployment rates, this may create expectations that there will be a timber 'shortage' and hence rising prices later on. If so, private sector timber producers may find it profitable to delay some of their cutting decisions.

In sum, the results derived in this subsection imply that:

Proposition 11.3
Unemployment (temporary or persistent) is not a sufficient condition to justify marginal departures from the Faustmann solution to the forest management problem.

11.5.2 *Keynesian Unemployment*

There are also important situations where one can reasonably expect that world timber prices are slow to adjust to eliminate world excess supply or demand. Or, there may be domestic laws regulating the permissible cut. In such situations one would expect either timber producers *or* domestic users of timber to face quantity constraints. This section is devoted to a brief examination of the case where domestic timber firms perceive a constraint on their final output, while households are unemployed. Such a situation is known as Keynesian unemployment in modern macroeconomic theory.

Suppose private timber firms can sell only $\bar{c}_1$ in period 1 and not the unconstrained profit maximizing quantity c_1^* ($>\bar{c}_1$) because demand is insufficient at prevailing prices. Furthermore, assume firms are unconstrained in period 2 so that they can sell at prevailing prices whatever quantity (including growth) is left from period 1. (For the properties of the timber supply function when a firm perceives quantity constraints, see Johansson and Löfgren, 1983b.)

In a situation with Keynesian unemployment, one would expect politicians to try to persuade the public timber producer to increase cutting and, hence, employment. However, without going into details, it can be shown that if the public firm succeeds in selling additional amounts of timber in period 1, this will crowd out private sector supply. This is because the public firm cannot affect aggregate demand, implying that *only a redistribution of sales and employment between firms occurs*. Private firms, which are now unconstrained in period 2, increase their sales while the public firm reduces its cutting (as it has already cut some of the trees which otherwise would be cut in period 2). Hence, aggregate employment in period 2 is left unchanged (unless firms differ in efficiency). This result indicates the need for policy makers to ascertain whether or not unemployment is due to deficiency of demand for output before adopting a policy that increases public sector employment. The reader interested in detailed cost–benefit rules (for ordinary produced goods) under Keynesian unemployment is referred to Cuddington, Johansson and Löfgren (1984, chapter 8) and Johansson (1982).

11.6 Cutting Rules in Excess Demand Situations

Another important rationing situation arises when domestic users of timber face a quantity constraint on their timber demand. In Western European countries like Finland and Sweden one can find strong empirical support for the view that the level of production of the forest industry is restricted by a 'shortage' of wood (see Johansson, 1981 and Johansson and Löfgren, 1983b). According to Bourchier (1982, p. 70) at least local wood shortages are observed in Canada.

If tradeables' producers (i.e. the forest industry) face a shortage of timber, they maximize (11.1) subject to the timber market constraint $c_{1t} = \bar{c}_{1t}$. The resulting first order conditions for profit maximization are

$$p_{1t}\frac{\partial f_{1t}}{\partial c_{1t}} - p_t - \beta_t = 0 \qquad \text{all } t \qquad p_{1t}\frac{\partial f_{1t}}{\partial l_{1t}} - w_t = 0 \tag{11.36}$$

where $\beta_t \geqslant 0$ is the Lagrange multiplier associated with the constraint in the tth period in the timber market (see figure 11.7).

Suppose the timber market constraint is binding only in period 1, $\beta_1 > 0$, $\beta_2 = \ldots = \beta_T = 0$, and that the economy is suffering from unemployment. As before all prices and wages are assumed to be constant over time. Using the same procedure as in earlier sections of this chapter, the monetary welfare change measure can be shown to be

$$\mathrm{d}W_c = \beta_1\,\mathrm{d}\bar{c}_{11} + (w - \alpha)\,\mathrm{d}l_{11} \tag{11.37}$$

In arriving at (11.37), it has been assumed that the public timber firm advances some of its cutting decisions to period 1 ($\mathrm{d}c_{g1} = \mathrm{d}\bar{c}_{11}$)

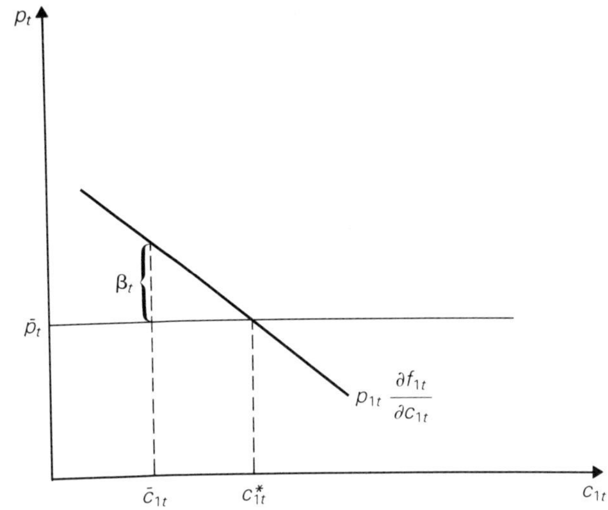

FIGURE 11.7 Rationing in the timber market. The firm faces a constraint $\bar{c}_{1t}$ on its timber demand, i.e. the firm cannot obtain all the timber it is prepared to buy at prevailing prices

but then returns to a Faustmann management of its forests. Hence $d\bar{\pi}_{g4} = 0$ as before. (See section 11.5.1.)

Private timber producers do not change their cutting decisions since all relative prices are fixed.

Thus we have the following rule for project evaluation when there is excess demand for timber and excess supply of labour:

Proposition 11.4

(a) Use the difference β_1 between the marginal willingness to pay for an additional unit of timber and the market price to evaluate rationed timber; and

(b) Determine the change in employment in the *forest industry* l_{11} and evaluate this change at the prevailing wage w less any adjustment α reflecting household disutility of increased work effort.

Observe that the sign of the change in employment in the forest industry is *a priori* uncertain. If timber and labour are technical complements (technical substitutes) employment increases (decreases) when rationing in the timber market is reduced. Inputs are technical complements (technical substitutes) if an increase in one input increases (decreases) the marginal product of the other input. Empirical observation suggests that timber and labour are technical complements so that employment increases. However, the forest industry uses other intermediate goods (produced by *labour* and other inputs) besides timber. *A priori* one cannot rule out the possibility that some of these intermediate goods and timber are technical substitutes. This means that the cost–benefit practitioner must in principle examine all inputs used by the forest industry in order to be able to calculate the employment term in the monetary change measure (11.37).

11.7 Concluding Remarks

The optimal management of a natural resource like forests is ultimately a long run issue and harvest decisions should be based on all present and future relative prices. At the same time, real world managers of forests have to base their decisions on an imperfect knowledge of e.g. long run price paths. Short run phenomena (like opposition to pollution, increasing unemployment rates in depressed areas of a country, and world excess demand for timber) often strongly affect the decisions. If this is admitted, it is an important task for economists to provide proper project evaluation rules for natural resources. Hopefully this chapter provides a reasonably realistic start to this enterprise.

Appendix: An Indirect Utility Function Approach

An alternative approach to that used in this chapter is to maximize an indirect utility function subject to certain equilibrium conditions. (The household's indirect utility function is analogous to the firm's present value function derived in chapter 6.) This approach yields shadow prices to be used by the public sector. In order to maximize welfare, a public firm should maximize profits calculated at these shadow prices. This can be shown to be equivalent to maximizing a monetary welfare (change) measure of the kind derived in this chapter, i.e. selecting such levels of public sector production that $dW = 0$ in e.g. (11.21).

To arrive at an indirect utility function, we first need the household's demand and supply functions. These functions, which are derived from the first order utility maximization conditions (11.17), can be written

$$\begin{aligned} q_{it}^{\mathrm{d}} &= q_{it}^{\mathrm{d}}(p_{11}, \ldots, w_T, \pi) \qquad & i = 1, \ldots, 3 \\ c_t^{\mathrm{d}} &= c_t^{\mathrm{d}}(p_{11}, \ldots, w_T, \pi) \qquad & \text{all } t \\ l_t^{\mathrm{s}} &= l_t^{\mathrm{s}}(p_{11}, \ldots, w_T, \pi) \end{aligned} \tag{11.38}$$

where

$$\pi = \sum_{j=1}^{4} (\pi_j + \pi_{gj})$$

is total profit income. Demand for goods and the supply of labour in any period are functions of all prices in all periods and the present value of the sum of distributed private sector and public sector profits.

Substituting (11.38) into the utility function (11.15) yields

$$\begin{aligned} V &= U(q_{11}^{\mathrm{d}}(p_{11}, \ldots, w_T, \pi), \ldots, l_T^{\mathrm{s}}(p_{11}, \ldots, w_T, \pi)) \\ &= V(p_{11}, \ldots, w_T, \pi) \end{aligned} \tag{11.39}$$

To examine the properties of this function, differentiate the function with respect to any price, say p_{it}. Using the first order conditions (11.17) for utility maximization, one obtains

$$\frac{\partial V}{\partial p_{it}} = \nu \left(p_{11} \frac{\partial q_{11}^{\mathrm{d}}}{\partial p_{it}} + \ldots + p_T \frac{\partial c_t^{\mathrm{d}}}{\partial p_{it}} - w_1 \frac{\partial l_1^{\mathrm{s}}}{\partial p_{it}} - \ldots - w_T \frac{\partial l_T^{\mathrm{s}}}{\partial p_{it}} \right) \tag{11.40}$$

However, by differentiating the budget constrain (11.16) with respect to p_{it}, it is seen that

$$-\left(p_{11} \frac{\partial q_{11}^{\mathrm{d}}}{\partial p_{it}} + \ldots + p_T \frac{\partial c_T^{\mathrm{d}}}{\partial p_{it}} - w_1 \frac{\partial l_1^{\mathrm{s}}}{\partial p_{it}} - \ldots - w_T \frac{\partial l_T^{\mathrm{s}}}{\partial p_{it}} \right) = q_{it}^{\mathrm{d}} \tag{11.41}$$

Hence, (11.40) reduces to

$$\frac{\partial V}{\partial p_{it}} = -\nu q_{it}^{\mathrm{d}}$$

In a similar way it can be shown that

$$\frac{\partial V}{\partial \pi} = \nu \tag{11.42}$$

where ν denotes the marginal utility of income.

Thus the indirect utility function

$$V = V(p_{11}, \ldots, w_T, \pi)$$

has the following properties:

$$\left.\begin{aligned}
\frac{\partial V}{\partial p_{it}} &= -\nu q_{it}^{\mathrm{d}}(p_{11}, \ldots, \pi) \\
\frac{\partial V}{\partial p_t} &= -\nu c_t^{\mathrm{d}}(p_{11}, \ldots, \pi) \\
\frac{\partial V}{\partial w_t} &= \nu l_t^{\mathrm{s}}(p_{11}, \ldots, \pi) \\
\frac{\partial V}{\partial \pi} &= \nu(p_{11}, \ldots, \pi)
\end{aligned}\right\} \quad \text{all } t \qquad (11.43)$$

The partial derivative of the indirect utility function with respect to the price of a commodity is equal to the optimal demand of that commodity multiplied by the marginal utility of income[18] (which, in turn, is equal to the partial derivative of the indirect utility function with respect to lump sum income).

To derive shadow prices to be used by the public sector, one maximizes the indirect utility function

$$V(p_{11}, \ldots, w_T, \pi) \qquad (11.44)$$

subject to the market clearing conditions

$$\begin{aligned}
&(\lambda_{p_{it}})\,(q_{it} + q_{git} - q_{it}^{\mathrm{d}} - q_{it}^{\mathrm{E}}) = 0 \qquad i = 1, 2, 3 \\
&(\lambda_{p_t})\,(c_t + c_{gt} - c_t^{\mathrm{d}} - c_t^{\mathrm{E}}) = 0 \\
&(\lambda_{w_t})\left[l_t^{\mathrm{s}} - \sum_{j=1}^{4} (l_{jt} + l_{gjt})\right] = 0 \qquad \text{all } t \qquad (11.45) \\
&(\lambda_B)\left(\sum_{i=1}^{3}\sum_{t=1}^{T} p_{it} q_{it}^{\mathrm{E}} + \sum_{t=1}^{T} p_t c_t^{\mathrm{E}}\right) = 0
\end{aligned}$$

where $\lambda_{p_{it}}$ etc. denote the Lagrange multipliers associated with the constraints, and a superscript E is used to denote net export of a commodity. Since the exchange rate is assumed to be fixed, the trade balance need not equal zero in each and every period. However, from

[18] The expression $(\partial V/\partial p_{it})/(\partial V/\partial \pi) = -q_{it}^{\mathrm{d}}$ is called Roy's identity.

Walras's law it follows that the sum over all periods of the current account in each period is equal to zero. This result is obtained by substituting for π in the household's budget constraint (11.16). This (in a sense redundant) balance of trade 'constraint' is captured by the last equation in (11.45).

Shadow prices can be derived by maximizing the Lagrange function – formed by (11.44) and (11.45) – with respect to q_{git}, c_{gt}, q_{it}^{E}, c_t^{E}, π and w_t (using the production functions and, for private firms, the first order conditions for profit maximization). Note that the goods prices p_{it} and p_t are world market determined owing to the small open economy assumption employed in this chapter and are, hence, exogenously determined.

It is left as an exercise to the reader to derive detailed shadow pricing rules. We will, however, indicate a few steps. First, maximization of the Lagrange function gives for the partial derivative with respect to the net export of the ith good in the tth period (q_{it}^{E})

$$-\lambda_{p_{it}} + \lambda_B p_{it} = 0 \tag{11.46}$$

Next, for the partial derivative with respect to the level of lump sum income (π), one obtains after straightforward calculations

$$\lambda_B = \nu \;(= \partial V/\partial \pi) \tag{11.47}$$

Hence $\lambda_{p_{it}}/\nu = p_{it}$, i.e. the shadow price of a commodity – normalized by the marginal utility of income – is equal to the market price of the commodity. It can be shown that this holds true for all goods and factors of the model.

From these derivations it should be rather obvious that maximization of a public firm's profit calculated at shadow prices is equivalent to maximization of the firm's profit calculated at market prices. In other words, in order to maximize welfare, a public firm should use the market prices as shadow prices. Compare with the monetary welfare change measures (11.21)–(11.25).

The indirect utility function approach can also be used to derive shadow prices to be used in the presence of 'market failures', e.g. unemployment and external effects. This generalization is not performed here. Instead the interested reader is referred to Marchand, Mintz and Pestieau (1983) and Roberts (1982).

12

The Properties of the Timber Supply Function in a Risky World

12.1 Introduction

Section 12.2 deals with a simplified stochastic variation of the rotation problem considered in chapters 4 and 5. The growth of the stand is assumed to be stochastic. Moreover, the forest owner is assumed to be able to monitor the development of the value of the stand.[1] The cutting rule which maximizes the expected present value of the cut is derived in this scenario. This rule, under fairly reasonable assumptions on the distribution function of growth, is analogous to the corresponding rule under certainty. The main text contains an intuitive proof, while the formal proof is available in the appendix to this chapter.

Sections 12.3 and 12.4 are devoted to a simplified version of the LNF model in chapter 6, which we use to investigate the behaviour of a risk averse forest owner who maximizes the present value of the cut. He is given the choice between cutting 'today' and/or cutting 'tomorrow', i.e. we collapse the multiperiod model of forestry into a two period intertemporal model. The problem of an uncertain future is highlighted by assuming that the net price of a cubic metre cut today is certain, while the present value of the net price tomorrow is a random variable. The general properties of a model of this kind are well known but, as far as we know, they have not yet been applied to forestry (see e.g. Hey, 1981, chapters 3, 4 and 8; Katz, Paroush and Kahana, 1982).

[1] An alternative would be to let the forest owner determine the optimal rotation period at time zero and force him to stick to this choice independently of the realizations (the outcomes) until the originally chosen cutting time. See Lippman and McCall (1980).

The results of this exercise are interesting. In a model where net prices are certain – as in chapter 6 – the properties of the timber supply function are very neoclassical. The vector of supplies is homogeneous of degree zero in net prices, i.e. a proportional tax levied on the present value of future revenues leaves the vector of optimal cuts unaffected. Moreover, fixed costs do not affect the optimal cut. This in turn means that the introduction of a lump sum tax, determined by for example the site quality classificaton index of the forest land, leaves the optimal cut unaffected. Finally, an increased price in a particular period increases the cut in that particular period, i.e. the supply curve is not backward bending.

None of these results holds true for the model presented in this chapter. Some of the six different results presented below are: the supply curve may be both forward bending or backward bending; the introduction of a proportional tax will increase the cut today and decrease the cut tomorrow; the same is true for the introduction of a lump sum tax; while an increase in the uncertainty of tomorrow will increase the cut today. To derive these results we have to assume, in addition to risk aversion, that he Arrow-Pratt index of absolute (relative) risk aversion is a decreasing function of the present value of the cut (Arrow, 1971; Pratt, 1964). Loosely speaking, this means that the higher the total profit, the more risks the forest owners are willing to take.

12.2 The Stochastic Tree

The particular analytical problem to be considered in this section is not a pure rotation problem[2] but rather a wine aging problem. Nevertheless, it illustrates important effects of stochastic elements in capital theory. The arguments introduced are borrowed from Brock, Rothschild and Stiglitz (1979) and Löfgren and Ranneby (1983), who extend the Brock, Rothschild and Stiglitz results to more general growth processes.

Assume that the value of the stand follows a discrete time Markov process

$$V_0, V_1, V_2, \ldots, V_t \tag{12.1}$$

with stationary transition probabilities $P(\cdot/V)$. If the price is constant we can regard V_t as the stocking level of the stand at time t. Alternatively the price and/or value may be stochastic.

[2] For how to handle a price rotation problem see Miller and Voltaire (1983).

If ϵ_t is growth, the conditional distribution of ϵ_t given $V_t = V$ is given by $P(\cdot/V)$. In the intuitive discussion, we assume that the growth increments are independent and identically distributed (IID) with mean μ. This means that the expected growth of the stand is independent of the age of the stand. The realism of this assumption for the growth of a forest stand is dubious, but can be relaxed without changing any essential properties of the optimal cutting rule.

Assume that the forest owner maximizes the expected discounted value of the cut:

$$\max_{T_\nu} E[V(T_\nu)\, \beta^{T_\nu}] \tag{12.2}$$

where T_ν = integer T, and $\beta = (1 + r)^{-1}$.

It is intuitively clear that the optimal rule under these simple conditions must be to choose a stand value (volume) $\bar{V}$ and first cut the stand when $V_t \geqslant \bar{V}$ (see also the appendix to this chapter). If the process is deterministic,

$$V_{t+1} = V_t + \mu \tag{12.3}$$

and the optimal cutting value (volume) must fulfil the following marginal conditions:

$$V = \beta(V + \mu) \tag{12.4}$$

The LHS represents the value of cutting the tree down now, while the RHS is the present value of the stand in the next period. In other words, the optimal cutting value has the property that the forest owner is indifferent between cutting now or letting the stand grow for another 'period'.[3] This yields the well known Wicksellian wine aging rule

$$\frac{\mu}{\bar{V}} = r \qquad \text{or} \qquad \bar{V} = \frac{\mu}{r} \tag{12.5}$$

i.e. the optimal t, and hence also the 'optimal stocking level' $\bar{V}$, are determined by equality between the interest rate and the rate of increase in the value of the stand.

Uncertainty does not influence the optimal stocking level (Brock, Rothschild and Stiglitz, 1979); assuming that the value (volume) of the stand is strictly increasing, $\text{prob}(\epsilon_t < 0) = P[(-\infty, V)/V] = 0$. Let $H(\hat{V})$ be the optimal expected value of a stand with volume $\hat{V}$. If $\hat{V}$ is the optimal cutting size,

$$H(V) = V \qquad \text{for} \quad V \geqslant \hat{V}$$

[3] The intuitive reasoning implicitly presupposes continuous time.

and

$$H(V) > V \qquad \text{for} \quad V < \hat{V} \tag{12.6}$$

i.e. if the actual stand value (stocking level) is larger than the optimal cutting level the stand should be cut and value (volume) equals the actual value (volume). If $V < \hat{V}$ the actual value (volume) is lower than the optimal and the value of having a stand of that size is larger than the actual value.

Moreover, at $\hat{V}$ the forest owner must be indifferent between cutting the stand down and letting it grow for another period, i.e.

$$\hat{V} = \beta E\{\max[\hat{V} + \epsilon, H(\hat{V} + \epsilon)]\} \tag{12.7}$$

However, since ϵ is non-negative with probability 1, $H(\hat{V} + \epsilon) = \hat{V} + \epsilon$ and (12.7) reduces to

$$\hat{V} = \beta E(\hat{V} + \epsilon) = \beta(\hat{V} + \mu) \quad \text{and} \quad \hat{V} = \frac{\mu}{r} \tag{12.8}$$

which we recognize from before; i.e. $\hat{V} = \bar{V}$ and the stand will be cut whenever $V \geqslant \hat{V}$.

If $P(\epsilon_t < 0) \neq 0$ the argument so far is no longer valid, since in this case $H(\hat{V} + \epsilon) > \hat{V} + \epsilon$ for $\epsilon < 0$, i.e. when left to grow another period and growth is negative, the optimal expected gain is larger than the actual value of the stand. A heuristic argument due to Brock, Rothschild and Stiglitz (1979) runs like this. Let $f(\epsilon)$ be the density function of ϵ with support on $[a, b]$ $(a < 0)$.[4] Then (12.8) becomes

$$\begin{aligned}
\hat{V} &= \left(\beta \int_a^0 H(\hat{V} + \epsilon) f(\epsilon)\, d\epsilon + \int_0^b (\hat{V} + \epsilon) f(\epsilon)\, d\epsilon\right) \\
&> \beta\left(\int_a^0 (\hat{V} + \epsilon) f(\epsilon)\, d\epsilon + \int_0^b (\hat{V} + \epsilon) f(\epsilon)\, d\epsilon\right) \\
&= \beta\Big(\hat{V} \underbrace{\int_a^b f(\epsilon)\, d\epsilon}_{1} + \underbrace{\int_a^b \epsilon f(\epsilon)\, d\epsilon}_{\mu}\Big) = \beta(\hat{V} + \mu)
\end{aligned}$$

which means that the optimal 'stopping value' (stocking level) exceeds the optimal value (stocking level) under certainty ($\hat{V} > \mu/r = \bar{V}$). We have shown:

[4] From the IID assumption it follows that $f(\epsilon)$ is independent of V.

Proposition 12.1
If the stationary transition probabilities are such that

$$P[(-\infty, V)/V] = 0$$

for all V (growth cannot be negative) and the expected growth function is IID, i.e.

$$\mu(V) = E(\epsilon_t/V_t = V) = \mu \text{ (constant) all } V$$

then the optimal policy π^* for a maximum level of expected present value is given by

$$\pi^*: \begin{cases} \text{if } V_t < \bar{V}, \text{ continue} \\ \text{if } V_t \geqslant \bar{V}, \text{ cut the stand} \end{cases}$$

where $\bar{V}$ is the optimal value of the stand in the deterministic case.

If $P[(-\infty, V)/V] > 0$ then the optimal stopping value exceeds $\bar{V}$.

Note that the optimal policy in the stochastic case does not provide any information about the *expected* optimal rotation period compared with the optimal rotation period in the deterministic case. It is shown in Lohmander (1984) that the relationship between the two rotation periods hinges in a rather complicated manner on the properties of the growth function. Proposition 12.1 rests on an assumption of independent and identically distributed growth increments. This assumption is unrealistic. In the appendix we show (by borrowing a proof due to Löfgren and Ranneby, 1983) that proposition 12.1 still holds if the growth function is generalized to

$$\mu(V) = E(\epsilon_t/V_t = V)$$

and $\mu(V)$ is a concave function. This is, except perhaps for the Markov property that value growth only depends on the actual value of the stand, a realistic description of timber growth.

12.3 The Maximization of Expected Utility and the Risk Averse Forest Owner

The objective of maximum expected value is frequently used in the literature dealing with economic decisions under risk. However, at times it yields hypotheses which are refuted, even by casual empiricism. For example, what is the maximum amount of money an expected value maximizing agent would be willing to pay to participate in a heads and tails game, where he is given 2^i dollars if heads turns up the ith time? The expected value of the game is

$$E(\pi) = \tfrac{1}{2}2 + (1-\tfrac{1}{2})\tfrac{1}{2}2^2 + (1-\tfrac{1}{2})(1-\tfrac{1}{2})\tfrac{1}{2}2^3 \ldots$$
$$= 1 + 1 + 1 + \ldots = \infty \qquad (12.9)$$

where the heads possibility is 1/2. The answer is an unlimited amount.[5]

A more general objective is to maximize the expected utility of profits:

$$E[U(\pi)] = \tfrac{1}{2}U(2) + \tfrac{1}{4}U(4) + \tfrac{1}{8}U(8) + \ldots \qquad (12.10)$$

If $dU/d\pi = U'(\pi) \geqslant 0$ and $d^2U/d\pi^2 < 0$ the series (12.10) might very well converge, meaning that the utility of participating in the game is finite. $U' \geqslant 0$ means that the marginal utility of further profits is non-negative, while $U'' < 0$ means that marginal utility is decreasing. The reader should note that $U'' = 0$ converts the maximization of expected utility of profit into a problem which is equivalent to the maximization of expected profit. $U'' = 0$ all π means that $U'(\pi) =$ constant, and the maximization of $E(a\pi)$, when $a > 0$, has the same solution as the maximization of $E(\pi)$ because $E(a\pi) = aE(\pi)$.

John von Neumann and Oscar Morgenstern (1947) founded this so-called expected utility doctrine. We here briefly introduce the Neumann-Morgenstern utility theory (see also Varian, 1978). This theory is concerned with individuals' choice among risky alternatives.

We begin the study of the behaviour under uncertainty by assuming that the forest owner has a (twice continuously differentiable) utility function

$$u(\pi);\; du/d\pi > 0 \qquad (12.11)$$

i.e. utility is assumed to be an increasing function of the level of profit π from cutting.

Suppose that the forest owner faces a cutting programme that yields one of n possible outcomes, i.e. there is uncertainty about the outcome. To simplify the exposition we consider only a finite number of discrete outcomes. Denote the profit of the ith outcome by π_i $(i = 1, \ldots, n)$. This outcome or level of profit occurs with prob-

[5] This game is originally due to the mathematician Nicolaus Bernoulli (1695–1726), one of the famous mathematicians in the Bernoulli family. His father's (Jean Bernoulli) brother Jacques Bernoulli wrote *Ars Conjectandi* (1713), the foundations of probability calculus. The game is often referred to as the St Petersburg paradox (or problem) (Nicolaus was a professor in St Petersburg).

ability P_i $(0 < P_i < 1)$. The expected profit of the cutting programme is

$$E(\pi) = \sum_{i=1}^{n} P_i \pi_i; \ \sum_{i=1}^{n} P_i = 1 \tag{12.12}$$

i.e. a weighted average of all possible profit levels. Similarly, the expected utility of the cutting programme can be written

$$E[u(\pi)] = \sum P_i u(\pi_i) \tag{12.13}$$

where the utility resulting from profit π_i is weighted by the probability that the forest owner ends up experiencing profit level π_i. Note that if the probability distribution is continuous, (12.12) and (12.13) transform into

$$E(\pi) = \int_a^b \pi g(\pi) \, \mathrm{d}\pi \tag{12.12a}$$

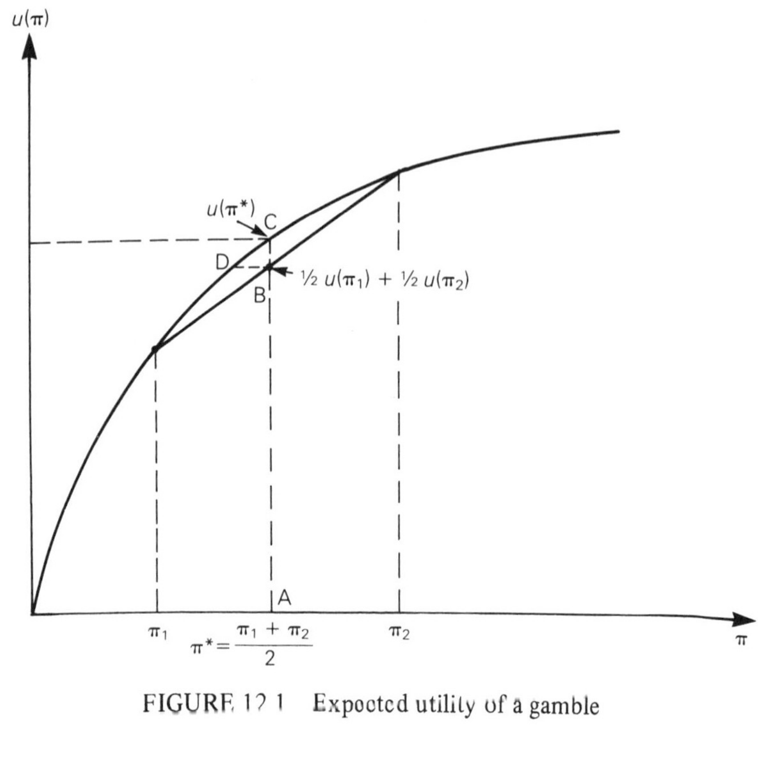

FIGURE 12.1 Expected utility of a gamble

$$E[u(\pi)] = \int_a^b u(\pi)\, g(\pi)\, d\pi \tag{12.13a}$$

where $g(\pi)$ is the density function of π with support on $[a, b]$.

Assume that the forest owner's utility function is strictly concave and has the shape depicted in figure 12.1, which is adapted from Varian (1978, p. 109), so that $du/d\pi > 0$, $d^2u/d\pi^2 < 0$. We make this assumption since the forest owner in section 12.4 is assumed to be risk averse. A *risk averter* by definition has a strictly concave utility function. In addition, assume that the cutting programme yields profit π_1 with probability 0.5, and profit π_2 with probability 0.5 ($P_3 = \ldots = P_n = 0$). Since the actual outcome is either π_1 or π_2, the expected profit is

$$0.5\pi_1 + 0.5\pi_2 = \pi^* \tag{12.14}$$

giving point A in figure 12.1. The corresponding expected utility, point B in the figure, is

$$0.5u(\pi_1) + 0.5u(\pi_2) \tag{12.15}$$

However, the forest owner clearly prefers to get the expected value π^* of the 'gamble' $u(\pi^*)$ rather than the 'gamble' itself, since

$$u(\pi^*) = u(0.5\pi_1 + 0.5\pi_2) > 0.5u(\pi_1) + 0.5u(\pi_2) \tag{12.16}$$

In the figure, the left hand side of this inequality is represented by point C while the right hand side corresponds to point B. Clearly, point C is connected with a higher level of utility than point B. Such behaviour, i.e. preferring the expected value of a 'gamble' to the 'gamble' itself, is called risk aversion. By moving horizontally to the left from point B towards point D it can be seen that a risk averter is a person who would be willing to forgo some part of the profit in order to change a random prospect into a certain one.

The figure may also be used to characterize a *risk neutral* forest owner. Such a forest owner has a linear utility function ($du/d\pi > 0$, $d^2u/d\pi^2 = 0$). Clearly the risk neutral forest owner will be indifferent between certain and risky prospects. Finally, for a *risk lover*, the utility function is convex ($du/d\pi > 0$, $d^2u/d\pi^2 > 0$) so that a straight line between profit levels π_1 and π_2 is above the corresponding segment of the utility function. Thus, he may prefer a risky prospect to a certain one even if the former gives a lower expected profit. $u[E(\pi)] > E\{u(\pi)\}[< E\{u(\pi)\}]$ when the utility function is strictly concave (convex) is a well-known mathematical theorem often referred to as Jensen's inequality.

12.4 The Model and the Results

We now turn to an expected utility maximizing forest owner who maximizes the expected utility of the present value of the cut.[6] We collapse the intertemporal problem into a two period problem in order to view it as a monitoring problem related to the monitoring problem discussed in section 12.2. The forest owner determines how much to cut today, given today's known price and a probability distribution of tomorrow's uncertain prices. However, because the model is a two period one, a given level of today's cut implicitly determines the level of tomorrow's cut.

Assumption 12.1
Consider a forest owner who chooses between cutting and selling 'today', c_1, and cutting and selling 'tomorrow', c_2. In implicit form the strictly concave and twice continuously differentiable production function (production possibility curve) is

$$F(c_1, c_2) = 0 \tag{12.17}$$

where c_i is the number of cubic metres cut in period i $(i = 1, 2)$. The biotechnology is therefore very similar to the more general technology of the LNF model in chapter 6, where the cut $c = (c_1, \ldots, c_T)$ is chosen in a compact convex set C. We solve (12.17) explicitly for c_2:

$$c_2 = f(c_1) \tag{12.18}$$

where $\mathrm{d}c_2/\mathrm{d}c_1 = f' < 0$, $\mathrm{d}^2c_2/\mathrm{d}c_1^2 = f'' < 0$.

The set of feasible cuts is depicted in figure 12.2.

The forest owner is assumed to be a price taker in all markets, and to have full information about today's net timber price[7] (stumpage) p_1. However, the present value of the net timber price of tomorrow is random and written as $\tilde{\beta}p_2$, where:

$$\tilde{\beta} = \beta + \gamma\epsilon;\ E(\epsilon) = 0, \sigma_\epsilon^2 = 1, P(\epsilon > -\beta/\gamma) = 1 \tag{12.19}$$

[6] This section is adapted from Johansson and Löfgren (1984a).

[7] This means that we assume that felling costs per cubic metre have been deducted. An explicit treatment of felling costs would change very little.

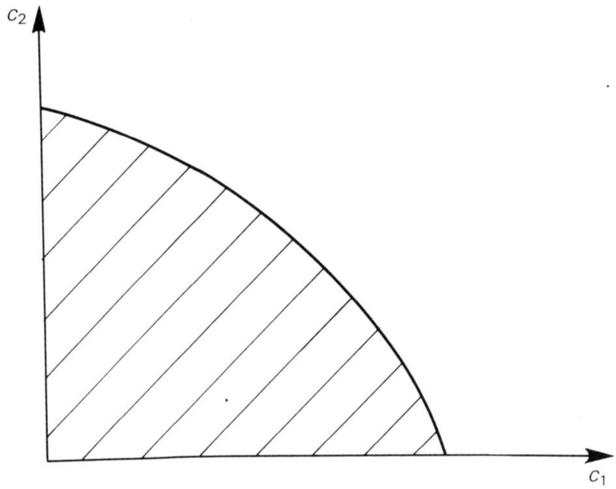

FIGURE 12.2 The set of feasible cuts, and the production possibility curve

Hence, $\tilde{\beta}(\epsilon)$ is a positive random variable, with distribution function $G(\epsilon)$ and density function $g(\epsilon) = G'(\epsilon)$, $E(\tilde{\beta}) = \beta$ its expected value, and γ its standard deviation. The last result follows by noting that $\sigma_\epsilon^2 = 1$ so that the variance of $\tilde{\beta}$ equals $E(\tilde{\beta} - \beta)^2 = E(\gamma\epsilon)^2 = \gamma^2 E\epsilon^2 = \gamma^2$. A spread preserving increase in tomorrow's price may be represented by an increase in β and an increase in γ may be used to represent a mean preserving increase in the variability of the price. The condition $P(\epsilon > -\beta/\gamma) = 1$ guarantees that tomorrow's price cannot be negative.

As indicated in section 12.3, the forest owner is a Neumann-Morgenstern utility maximizer with utility function $u(\pi)$ defined on the present value of profits π. The problem of the forest owner is to choose c_1 to maximize expected utility of profits:[8]

$$V = E\{u(\pi)\} = E\{u[p_1 c_1 + \tilde{\beta} p_2 f(c_1)]\} = \int_a^b u[\pi(\epsilon)]\, g(\epsilon)\, d\epsilon \tag{12.20}$$

where $[a, b]$ is the support of $g(\epsilon)$. Throughout the rest of this chapter we will assume that the forest owner is a risk averter, i.e. that $u''(\pi) < 0$.

[8] Throughout it is assumed that the forest owner must decide upon the level of cutting before the uncertainty about tomorrow's net price is resolved.

The first order and the second order conditions for utility maximization are obtained from (12.20):

$$\frac{\partial V}{\partial c_1} = V_1 = \int_a^b u'(\pi)\frac{\partial \pi}{\partial c_1} g(\epsilon)\,\mathrm{d}\epsilon$$

$$= \int_a^b u'(\pi)\,[p_1 + \tilde{\beta} p_2 f'(c_1)]\, g(\epsilon)\,\mathrm{d}\epsilon$$

$$= E[u'(p_1 + \tilde{\beta} p_2 f')] = E[u'(p_1 + \beta p_2 f')] + \gamma p_2 f' E[u'\epsilon] = 0 \qquad (12.21)$$

$$\frac{\partial^2 V}{\partial c_1^2} = V_{11}$$

$$= \int_a^b \{u''(\pi)\,[p_1 + \tilde{\beta} p_2 f'(c_1)]^2 + u'(\pi)\,\beta p_2 f''(c_1)\}\, g(\epsilon)\,\mathrm{d}\epsilon$$

$$= E[u''(p_1 + \tilde{\beta} p_2 f')^2] + E[u' \tilde{\beta} p_2 f''] < 0 \qquad (12.22)$$

This solution may be compared to that obtained under certainty (given assumptions 12.1). In the latter case $\tilde{\beta} = \beta$, so that the first order condition reads

$$p_1 + \beta p_2 f'(c_1) = 0 \qquad (12.23)$$

i.e. the net revenue of cutting one more cubic metre today equals the net revenue forgone tomorrow. The sign of the corresponding expression under uncertainty depends, according to (12.21), on the sign of $\gamma p_2 f' E[u'\epsilon]$. By definition $\gamma p_2 > 0$ and $f' < 0$. The term $E[u'\epsilon]$ expresses $\mathrm{cov}(u', \epsilon)$. This is because

$$\mathrm{cov}\, u'(\pi)\,\epsilon = \int_a^b \{[u'(\pi) - u'_e]\,(\epsilon - 0)\}\, g(\epsilon)\,\mathrm{d}\epsilon$$

$$= \int_a^b u'(\pi)\,\epsilon g(\epsilon)\,\mathrm{d}\epsilon - u'_e \int_a^b g(\epsilon)\,\mathrm{d}\epsilon$$

$$= \int_a^b u'(\pi)\,\epsilon g(\epsilon)\,\mathrm{d}\epsilon = E[u'\epsilon] \qquad (12.24)$$

where $u'_e = Eu'$. To sign $\mathrm{cov}(u'\epsilon)$ we take the partial derivative of $u'(\pi)$ with respect to ϵ to obtain

$$\partial u'(\pi)/\partial \epsilon = \gamma p_2 f(c_1)\, u'' < 0 \qquad (12.24a)$$

since $u'' < 0$. Thus, $E[u'\epsilon] < 0$ (and $\gamma p_2 f' E[u'\epsilon] > 0$).[9] Therefore, the expression $p_1 + \beta p_2 f'$ must be negative in (12.21). Since $p_1 + \beta_2 p_2 f'$ is a decreasing function of c_1, this implies the following proposition:

Proposition 12.2
Given assumptions 12.1, a risk averting forest owner will cut more today and less tomorrow when there is uncertainty in tomorrow's market than when there is no uncertainty.

Next, consider an increase in today's net timber price p_1. Differentiating (12.21) with respect to p_1 gives

$$V_{11}\,\mathrm{d}c_1 + \frac{\mathrm{d}}{\mathrm{d}p_1}\int_a^b \{u'(\pi)[p_1 + \beta p_2 f'(c_1)]\, g(\epsilon)\,\mathrm{d}\epsilon\}\,\mathrm{d}p_1 = 0 \quad (12.25)$$

or

$$\partial c_1/\partial p_1 = -[E\{u''(p_1 + \tilde{\beta} p_2 f')\, c_1\} + Eu']/V_{11} \quad (12.25a)$$

The denominator of this expression is negative from (12.22) while $Eu' > 0$. To determine the sign of the remaining term in the numerator we use what is known as the Arrow-Pratt index of absolute risk aversion $R(\pi)$:

$$R(\pi) = -u''(\pi)/u'(\pi) \quad (12.26)$$

In section 12.3 the sign of u'' is an indicator of risk attitudes. However, if we multiply the utility function by say 2, u'' changes to $2u''$ although behaviour is unaffected since the utility function is unique up to a linear transformation. This is the reason for 'normalizing' by u'. In the appendix to this chapter, it is shown that:

Assumption 12.2
If the forest owner becomes less averse to risk as profit increases, $R' < 0$, then

$$E[u''(p_1 + \tilde{\beta} p_2 f')] < 0 \quad (12.27)$$

Absolute risk aversion.

[9] $$\operatorname{cov} u'\epsilon = \int_a^b u'(\pi)\,\epsilon g(\epsilon)\,\mathrm{d}\epsilon = \int_a^b \left[u'(0) + \frac{\partial u}{\partial \epsilon}(\epsilon_0)(\epsilon - 0)\right] \epsilon g(\epsilon)\,\mathrm{d}\epsilon$$

where ϵ_0 belongs to the open interval $(0, \epsilon)$. This expression simplifies to

$$\operatorname{cov} u'\epsilon = 0 + \int_a^b g(\epsilon)\,\epsilon^2 \frac{\partial u}{\partial \epsilon}(\epsilon_0)\,\mathrm{d}\epsilon < 0$$

The effects on the levels of cutting from an increase in today's net timber price are summed up in the following proposition:

Proposition 12.3
Given assumptions 12.1 and no uncertainty, an increase in today's net timber price p_1 will increase cutting today and decrease future cutting.[10] However, if there is uncertainty in tomorrow's market and the forest owner has decreasing absolute risk aversion (assumption 12.2) an increase in p_1 will affect cutting levels in an ambiguous way.

The reason for this rather counter-intuitive result under uncertainty can be explained in the following way. Firstly, there is a 'substitution' effect of the increase in today's net price of timber. This effect, which is the only effect under certainty, stimulates today's cutting and decreases tomorrow's cutting. Secondly, the price increase makes the forest owner richer. By assumption [see (12.27)], this induces the forest owner to take more risks, i.e. induces him to 'wait' and cut more tomorrow. The sign of the sum of these counteracting effects is ambiguous.

This 'inelasticity' result is not inconsistent with 'reality'. Roundwood prices in Sweden have increased over the last few years to reduce excess demand in this market. There has, however, only been a slight effect on supply from changing prices. Sometimes this is explained by a backward bending supply curve, owing to income constraints. The model used in this chapter provides another, and perhaps more probable, explanation. As shown in chapter 7, there are strong theoretical reasons for rejecting the hypothesis that the deterministic supply curve is backward bending. The estimations in chapter 9 support this theoretical conclusion.

There are two ways of examining the effects of changes in the distribution of tomorrow's net timber price. Let us first consider a spread preserving increase, i.e. a change in β. Differentiating (12.21) with respect to β one obtains

$$\partial c/\partial\beta = -[E\{u''(p_1+\tilde{\beta}p_2 f')\,p_2 f(c_1)\} + E\{u'p_2 f'\}]/V_{11} < 0 \tag{12.28}$$

since the denominator and both terms in the numerator are negative. Thus:

[10] This has not been shown but should be fairly obvious from (12.23), figure 12.2 and the treatment in chapter 6.

Proposition 12.4
Given assumptions 12.1 and 12.2, a spread preserving increase in tomorrow's net timber price, $d\beta > 0$, will decrease today's cutting and increase tomorrow's cutting. An increase in β will give the same qualitative result under certainty.[11]

Next, consider a mean preserving increase in the variability of tomorrow's net timber price. From (12.21),

$$\partial c_1/\partial\gamma = -\{E(u''(p_1 + \tilde{\beta}p_2 f')[p_2 f(c_1)\epsilon]) + p_2 f' E(u'\epsilon)\}/V_{11} \qquad (12.29)$$

The second term of the numerator is positive from (12.24). As shown in the appendix to this chapter, the first term is also positive, so that $\partial c_1/\partial\gamma > 0$, i.e.:

Proposition 12.5
Given assumptions 12.1 and 12.2 a mean preserving increase in the variability of tomorrow's net timber price will increase (decrease) today's (tomorrow's) cutting.

The economic rationale behind this result is clear. The increased uncertainty regarding tomorrow's price combined with decreasing absolute risk aversion makes the forest owner less motivated in taking risks. Therefore, he shifts his cutting towards the present.

There are two additional changes of obvious economic interest to consider – a change in the fixed part of total cutting costs (a 'Finnish lump sum tax') and a change in profit taxes. The comparative statics effect of an increase in lump sum taxation T is, from (12.20) and (12.21),

$$\partial c_1/\partial T = E[u''(p_1 + \tilde{\beta}p_2 f')]/V_{11} > 0 \qquad (12.30)$$

which means that:

Proposition 12.6
Given assumptions 12.1 and 12.2, a lump sum tax will increase (decrease) today's (tomorrow's) cutting if tomorrow's net timber price is uncertain. On the other hand, if there is no uncertainty, a lump sum tax will leave cutting decisions unchanged.[12]

The lump sum tax decreases profits. Since the forest owner has decreasing absolute risk aversion, the reduction in profit makes him

[11] This follows from differentiation of (12.23).
[12] See the analyses in chapters 5 and 6, or note that $p_1 + \tilde{\beta}p_2 f' = p_1 + \beta p_2 f' = 0$ from the first order condition under certainty.

less interested in taking risks. Thus, he cuts more today and less tomorrow. If there is no uncertainty, the relative revenues from cutting today and tomorrow remain unchanged, implying that the forest owner has no incentive to change his cutting decisions. (The same holds true for a risk neutral person, i.e. a person for whom $U'' = 0$, in an uncertain world.)

Assume next that the forest owner must pay a proportional tax t on profit. His disposable profit then equals $(1-t)\pi$. The comparative statics effect of a change in t is

$$\partial c_1/\partial t = \{E(u''(p_1 + \tilde{\beta}p_2 f')[p_1c_1 + \tilde{\beta}p_2 f(c_1)])(1-t) + E[u'(p_1 + \tilde{\beta}p_2 f')]\}/V_{11}(1-t) \quad (12.31)$$

The second term of the numerator is clearly zero from (12.21). The Arrow-Pratt index of relative risk aversion

$$R_d(\pi_d) = -\pi_d u''(\pi_d)/u'(\pi_d) \quad (12.32)$$

where $\pi_d = \pi(1-t) > 0$, determines the sign[13] of the first term in (12.31). If relative risk aversion is decreasing ($R'_d < 0$), then

$$E[u''(p_1 + \tilde{\beta}p_2 f')\,\pi_d] < 0 \quad (12.33)$$

The proof is omitted since it parallels the proof of the corresponding expression for decreasing absolute risk aversion (see also Sandmo, 1971).

In sum:

Proposition 12.7

Given assumptions 12.1 and R'_d, an increased proportional taxation of profit increases (decreases) cutting today (tomorrow) in an uncertain world. If the uncertainty of tomorrow's price is eliminated, the change in the tax rate leaves all cutting decisions unchanged (see also chapters 5 and 6).

Once again the result under uncertainty is in sharp contrast with the 'conventional wisdom' whereby a proportional profit tax leaves all

[13] As Katz (1983) has pointed out, this definition of R_d is incompatible with decreasing or constant relative risk aversion when $\pi_d < 0$. He suggests that measures of risk aversion are properly defined upon wealth and not upon profits. Since wealth cannot in general be negative, the problem disappears. However, under the new definition (12.33) is no longer possible to sign under decreasing relative risk aversion.

real variables unchanged. The logic behind the result under uncertainty is the same as that discussed in connection with the lump sum tax. Given decreasing relative risk aversion, the forest owner will be less induced to take risks when the tax rate is increased.

Finally, it may be noted that the absence of real effects of lump sum taxes and profit taxes in a certain world has nothing to do with the slope of the supply curve. Thus, taxes of the kind considered will be neutral even if the supply curve is backward bending (provided $\pi_d > 0$). The model presented here offers a rather different explanation to the occasionally observed 'perverse' reactions of forest owners to proposed supply stimulating measures.

12.5 Concluding Comments

The realism of the model in section 12.4 can only be judged by testing its empirically meaningful theorems against reality. We feel that the hypotheses of the simple model of this chapter compare well with a pure neoclassical model of a certain world. Forestry involves long run decisions connected with uncertainty regarding future revenues, and economic models which do not recognize this feature are likely to predict poorly.

Appendix: Proofs of Claims

A Generalization of Proposition 12.1

Assume that the value of the stand follows a discrete time Markov process

$$V_0, V_1, V_2, \ldots, V_t \tag{12.34}$$

with stationary transition probabilities $P(A/x)$. If the price is constant, V_t may be regarded as the stocking level of the stand at time t. If ϵ_t denotes growth, the conditional distribution of ϵ_t given $V_t = x$ is given by $P(\cdot/x)$.

Assume that the forest owners maximize the expected discounted value of the cut and let $H(x)$ be the expected optimal gain function, i.e.

$$H(x) = \sup_{t \in N} E(V_t \beta^t / V_0 = x) \tag{12.35}$$

where N denotes the set of non-negative integers, $\beta = (1 + r)^{-1}$ is the discount factor, and $r > 0$ is the interest rate.

We assume that the state space of the process $\{V_t: t = 0, 1, 2 \ldots\}$ is a bounded subset of the real line. From the theory of Markov decision processes (see e.g. Ross, 1970), it follows that the optimal gain function $H(x)$ satisfies the following functional equation:

$$H(x) = \max\left[x, \beta \int H(y)\, P(\mathrm{d}y/x)\right] \tag{12.36}$$

where $\beta \int H(y)\, P(\mathrm{d}y/x)$ is the present value of the optimal policy provided that the stand is saved if in state x. Furthermore,

$$\lim_{n \to \infty} H(x, n) = H(x) \tag{12.37}$$

where $H(x, n)$ denotes the optimal gain function for an n period problem. We are now ready to prove the following claims:

Proposition 12.8
If the stationary transition probabilities are such that

$$P[(-\infty, x)/x] = 0$$

for all x (growth cannot be negative) and the expected growth function

$$\mu(x) = E(\epsilon_t / V_t = x)$$

is strictly concave,[14] then the optimal policy π^* is given by

$$\pi^*: \begin{cases} \text{if } V_t < x^*, \text{ continue} \\ \text{if } V_t \geqq x^*, \text{ cut the stand} \end{cases}$$

For the definition of x^* see figure 12.3.
The policy π^* is also optimal in the deterministic case.

Proof
Let B be the set of states where cutting is at least as good as leaving the stand to grow exactly one more period and then cutting, i.e.

$$B = \{x: x \geqslant \beta \int yP(\mathrm{d}y/x)\}$$

[14] As should be obvious from figure 12.3, the assumption of a strictly concave growth function can be considerably weakened. If β is close to one (r less than say 0.2) the set B, defined in the proof, will be connected (x^* unique) even if the growth function has a strictly convex segment for young stands (stands with low stocking levels). Hence, the optimal policy rule will hold for every growth function and every reasonable discount factor.

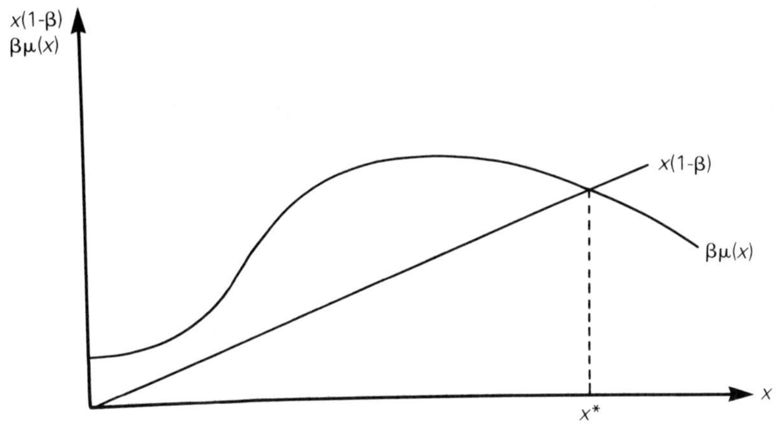

FIGURE 12.3 Determination of the set B

Since

$$\int yP(\mathrm{d}y/x) = x + \int (y-x)\,P(\mathrm{d}y/x) = x + \mu(x)$$

then

$$x \geqslant \beta \int P(\mathrm{d}y/x)$$

iff

$$x(1-\beta) \geqslant \beta\mu(x)$$

From figure 12.3 and this formula, it is obvious that

$$B = \{x \colon x \geqslant x^*\}$$

We now show that $H(x, n) = x$ for all $x \in B$ and all n. The definition of the set B implies that $H(x, 1) = x$ for all $x \in B$. Suppose that $H(x, n-1) = x$ for all $x \in B$. Clearly

$$H(x, n) = \max\left[x, \beta \int H(y, n-1)\,P(\mathrm{d}y/x)\right]$$

Since growth cannot be negative, once the process is in B it can never leave this set. Thus for $x \in B$

$$\beta \int H(y, n-1)\,P(\mathrm{d}y/x) = \beta \int_B H(y, n-1)\,P(\mathrm{d}y/x)$$

$$= \beta \int_B yP(\mathrm{d}y/x) = \beta \int yP(\mathrm{d}y/x) \leqq x$$

implying that $H(x, n) = x$ for all $x \in B$.

By letting $n \to \infty$ we obtain

$$H(x) = x \quad \text{for } x \in B$$

Now, for $x \notin B$, the policy which continues for exactly one stage and then stops (the stand is cut) has the expected gain

$$\beta \int yP(\mathrm{d}y/x)$$

which is strictly greater than x (since $x \notin B$). Thus the optimal gain function $H(x)$ is such that

$$H(x) = \begin{cases} x \text{ if } x \in B, \text{ i.e. } x \geqslant x^* \\ > x \text{ if } x \notin B, \text{ i.e. } x < x^* \end{cases}$$

and the optimal policy is given by π^*. If growth is deterministic, the transition probabilities are given by $P[x + \mu(x)/x] = 1$ satisfying $P[(-\infty, x)/x] = 0$, and the optimal policy is still given by π^*.

What happens if growth (value growth) is negative but still has expected growth $\mu(x)$? Since the optimal gain function $H(x)$ satisfies $H(x) \geqslant x$ for all x, we obtain

$$x \in B^c = \{x\colon x < x^*\} = \{x\colon x < \beta \int yP(\mathrm{d}y/x)\} \tag{12.38}$$

so that

$$\beta \int H(y)\, P(\mathrm{d}y/x) \geqq \beta \int yP(\mathrm{d}y/x) > x$$

Thus for $x < x^*$ it is optimal to leave the stand to grow at least another period. Consequently, the optimal stocking level under uncertainty can never be lower than the optimal stocking level under certainty.

The Implications of Decreasing Risk Aversion

To show (12.27), define π^0 as the level of profit obtained when $p_1 + \beta p_2 f' = 0$. Next, define $\pi > \pi^0$ as the profit corresponding to $\tilde{\beta} > \beta$. Then the Arrow-Pratt index gives

$$-u''(\pi)/u'(\pi) < -u''(\pi^0)/u'(\pi^0) \tag{12.39}$$

if absolute risk aversion is decreasing. Multiplying both sides of (12.39) by $p_1 + p_2\tilde{\beta} f' < 0$ and rearranging gives

$$u''(\pi)\,(p_1 + p_2\tilde{\beta} f') < [u''(\pi^0)/u'(\pi^0)]\, u'(\pi)\,(p_1 + \tilde{\beta} p_2 f') \tag{12.40}$$

If $\tilde{\beta} < \beta$ then $\pi < \pi^0$, so that the inequality (12.39) is reversed. However, as $p_1 + p_2\beta f' > 0$ when $\tilde{\beta} < \beta$, the inequality (12.40) still holds. Taking expectations, and using the first order condition (12.21) to show that the expected value of the right hand side of (12.40) is zero, it follows that

$$E[u''(p_1 + p_2\tilde{\beta}f')] \leqslant 0 \tag{12.41}$$

[If $\pi = \pi^0$ then $E\{u''(p_1 + p_2\beta f')\} = 0$.]

Next, we sign the first term of (12.29):

$$E\{u''(p_1 + \tilde{\beta}p_2 f')[p_2 f(c_1)\,\epsilon]\} \tag{12.42}$$

First, note that ϵ may be written

$$\epsilon = (1/\gamma)(\tilde{\beta} - \beta) \tag{12.43}$$

Substitution of (12.43) into (12.42) yields

$$\begin{aligned}[f(c_1)\,E\{u''(p_1 + \tilde{\beta}p_2 f')(\tilde{\beta}p_2 f' - \beta p_2 f' + p_1 - p_1)]/f'\gamma\} \\ = f(c_1)[E\{u''(p_1 + \tilde{\beta}p_2 f')^2\} - (p_1 + \beta p_2 f') \\ \times E\{u''(p_1 + \tilde{\beta}p_2 f')\}]/f'\gamma\end{aligned} \tag{12.44}$$

The first term within [ ] is negative as $Eu'' < 0$, while $(p_1 + \beta p_2 f') < 0$ from (12.21) and $E[u''(p_1 + \tilde{\beta}p_2 f')] < 0$ from (12.41). Since $f(c_1)/f'\gamma$ clearly is negative, it follows that the whole expression, and therefore also (12.42), is positive.

Bibliography

Adams, D. M. (1977) *Effects of National Forest Timber Harvest on Softwood Stumpage, Lumber and Plywood Markets: An Econometric Analysis*, Oregon State University, Forest Laboratory, research bulletin 15.

Arrow, K. J. (1951) *Social Choice and Individual Values* (New York: John Wiley & Sons).

Arrow, K. J. (1959) Towards a Theory of Price Adjustment, in Abramowitz, M. (ed.) *The Allocation of Economic Resources* (Stanford: Stanford University Press).

Arrow, K. J. (1971) *Essays in the Theory of Risk Bearing* (Chicago: Markham).

Arrow, K. J. and Debreu, G. (1954) Existence of Equilibrium for a Competitive Economy, *Econometrica*, **22**, 522–52.

Arrow, K. J. and Hahn, F. H. (1971) *General Competitive Analysis* (San Francisco: Holden-Day).

Arrow, K. J. and Kurz, M. (1970) *Public Investment, the Rate of Return and Optimal Fiscal Policy* (Baltimore: Johns Hopkins University Press).

Balinski, M. L. and Baumol, W. J. (1968) The Dual in Nonlinear Programming and its Economic Interpretation, *Review of Economic Studies*, **35**, 237–56.

Barnett, H. J. and Morse, C. (1963) *Scarcity and Growth: The Economics of Natural Resource Activity* (Baltimore: Johns Hopkins University Press).

Baumol, W. J. and Oates, W. E. (1975) *The Theory of Environmental Policy* (Englewood Cliffs: Prentice-Hall).

Beddington, J. R., Watts, C. M. K. and Wright, W. D. C. (1975) Optimal Cropping of Self-reproducible Natural Resources, *Econometrica*, **43**, 789–802.

Bellman, R. (1957) *Dynamic Programming* (Princeton: Princeton University Press).

Berck, P. (1976) *Natural Resources in a Competitive Economy*, MIT, Ph.D. dissertation.

Berck, P. (1981) Optimal Management of Renewable Resources with Growing Demand and Stock Externalities, *Journal of Environmental Economics and Management*, **8**, 105–17.

Binkley, C. S. (1979) *Timber Supply from Private Nonindustrial Forests: An Econometric Analysis of Landowner Behavior*, Yale University, Ph.D. dissertation.

Blume, L., Rubinfeld, D. L. and Shapiro, P. (1984) The Taking of Land: When Should Compensation be Paid?, *Quarterly Journal of Economics*, **99**, 71–92.

Boadway, R. W. (1974) The Welfare Foundations of Cost–Benefit Analysis, *Economic Journal*, **84**, 926–39.

Boadway, R. W. (1975) Cost–Benefit Rules in General Equilibrium, *Review of Economic Studies*, **42**, 361–73.

Boadway, R. W. (1978) A Note on the Treatment of Foreign Reserves in Project Evaluation, *Economica*, **45**, 391–9.

Boadway, R. W. and Bruce, N. (1984) *Welfare Economics* (Oxford: Basil Blackwell).

Boiteux, M. (1956) Sur la Gestion des Monopoles Publics Astreints à l'Equilibre Budgétaire, *Econometrica*, **24**, 22–40.

Boulding, K. E. (1935) The Theory of a Single Investment *Quarterly Journal of Economics*, **49**, 475–94.

Bourchier, R. J. (1982) Forestry in Canada: National Needs but Provincial Control, in Hewett, C. E. and Hamilton, T. E. (eds) *Forests in Demand* (Boston: Auburn House Publishing Company).

Brännlund, R., Johansson, P. O. and Löfgren, K. G. (1983) *An Econometric Analysis of Timber Supply in Sweden*, Department of Forest Economics, Swedish University of Agricultural Sciences, report no. 44. Revised version forthcoming in *Forest Science*.

Brännman, L. (1984) The Effects of Forest Service Hauling Distances on Dependent Communities and Winning Bids Forthcoming in *Land Economics*.

Brock, W., Rothschild, M. and Stiglitz, J. (1979) *Stochastic Capital Theory*, Department of Economics, University of Wisconsin, Madison, mimeographed.

Brown, G. Jr (1977) Promising Research Topics Regarding Non-timber Products from Forest Related Lands, in Clawson, M. (ed.) *Research in Forest Economics and Forest Policy* (Washington, DC: Resources for the Future).

Calish, S., Fight R. D. and Teeguarden, D. E. (1978) How Do Nontimber Values Affect Douglas-Fir Rotations?, *Journal of Forestry*, **76**, 217–21.

Cantor, D. G. and Lippman, S. A. (1983) Investment Selection with Imperfect Capital Markets, *Econometrica*, **51**, 1121–44.

Clark, C. W. (1976) *Mathematical Bioeconomics: The Optimal Management of Renewable Resources* (New York: John Wiley & Sons).

Clark, C. W., Clark, F. H. and Munro, G. R. (1979). The Optimal Exploitation of Renewable Resource Stocks: Problems of Irreversible Investment, *Econometrica*, **47**, 25–47.

Clark. C. W. and Munro, G. R. (1975) Economics of Fishing and Modern Capital Theory: A Simplified Approach, *Journal of Environmental Economics and Management*, **2**, 92–106.

Clawson, M. and Knetsch, J. L. (1966) *Economics of Outdoor Recreation* (Baltimore: Johns Hopkins University Press).

Coase, R. H. (1960) The Problem of Social Cost, *Journal of Law and Economics*, **3**, 1–44.

Crutchfield, J. A. and Zellner A. (1962) Economic Aspects of the Pacific Halibut Fishery, *Fishery Industrial Research*, vol. 1, no. 1 (Washington: US Department of the Interior).

Cuddington, J. T. Johansson, P. O. and Löfgren, K. G. (1984) *Disequilibrium Macroeconomics in Open Economies* (Oxford: Basil Blackwell).

Dasgupta P. S. (1982) *The Control of Resources* (Oxford: Basil Blackwell).

Dasgupta, P. S. and Heal, G. M. (1979) *Economic Theory and Exhaustible Resources* (Oxford: James Nisbet/Cambridge University Press).

Debreu, G. (1959) *Theory of Value* (New York: John Wiley & Sons).

Devarajan, S. and Fisher, A. C. (1981) Hotelling on Exhaustible Resources, *Journal of Economic Literature*, **19**, 65–73.

Diamond, P. A. and Mirrlees J. A. (1971) Optimal Taxation and Public Production II, *American Economic Review*, **61**, 261–78.

Dorfman, R. (1969) An Economic Interpretation of Optimal Control Theory, *American Economic Review*, **59**, 817–31.

Drèze, J. H. (1974) Discount Rates and Public Investment: A Postscriptum, *Economica*, **41**, 52–61.

Ekelund, R. B. J., Higgins, R. S. and Smithson, C. W. (1981) Can Discrimination Increase Employment: A Neo-classical Perspective, *Southern Journal of Economics*, **48**, 664–73.

Eriksson, H. (1976) *Granens Produktion i Sverige* (Stockholm: College of Forestry).

Faustmann, M. (1849) Berechnung des Wertes welchen Waldboden sowie noch nicht haubare Holzbestände für die Waldwirtschaft besitzen, *Allgemeine Forst und Jagd-Zeitung*, **25**, 441–55.

Fisher, A. C. (1981) *Resource and Environmental Economics* (Cambridge: Cambridge University Press).

Fisher, A. C. and Hanemann, M. W. (1983) *Endangered Species: The Economics of Irreversible Damage*, University of California, Berkeley, Department of Agricultural and Research Economics, working paper no. 329.

Formby, J. P. Layson, S. K. and Smith, W. J. (1983) Price Discrimination, Adjusted Concavity, and Output Changes under Conditions of Constant Elasticity, *Economic Journal*, **93**, 892–9.

Gordon, H. S. (1954) Economic Theory of Common Property Resources, *Journal of Political Economy*, **62**, 124–42.

Gregory, G. R. (1972) *Forest Resource Economics* (New York: John Wiley & Sons).

Hägglund, B. (1981) *Forecasting Growth and Yield in Established Forests*, Swedish University of Agricultural Sciences, Department of Forest Surveying, report no. 31.

Hanemann, M. W. (1984) *On Reconciling Different Concepts of Option Value*, University of California, Berkeley, Department of Agricultural and Resource Economics, mimeographed.

Hannesson, R. (1978) *Economics of Fisheries* (Oslo: Universitetsförlaget).

Harberger, A. C. (1971) Three Basic Postulates for Applied Welfare Economics: An Interpretative Essay, *Journal of Economic Literature*, **9**, 785–97.

Hartman, R. (1976) The Harvesting Decision where a Standing Forest has Value, *Economic Inquiry*, **14**, 52–8.

Heaps, T. (1981) The Qualitative Theory of Optimal Rotations, *Canadian Journal of Economics*, **14**, 686–99.

Heaps, T. (1984) The Forestry Maximum Principle, *Journal of Economic Dynamics and Control*, **7**, 131–51.

Heaps, T. and Neher P. A. (1979) The Economics of Forestry when the Rate of Harvest is Constrained, *Journal of Environmental Economics and Management*, **6**, 297–316.

Henderson, J. M. and Quandt, R. E. (1980) *Microeconomic Theory: A Mathematical Approach*, 2nd edn (Tokyo: McGraw-Hill).

Herfindahl, O. C. and Kneese, A. V. (1974) *Economic Theory of Natural Resources* (Columbus: C. E. Merrill Publishing Company).

Hey, J. D. (1981) *Economics in Disequilibrium* (Oxford: Martin Robertson).

Hirshleifer, J. (1970) *Investment, Interest and Capital* (London: Prentice-Hall).

Hoel, M. (1978) Extermination of Self-reproducible Natural Resources under Competitive Conditions, *Econometrica*, **46**, 219–24.

Hotelling, H. (1931) The Economics of Exhaustible Resources, *Journal of Political Economy*, **39**, 137–75.

Hotelling, H. (1932) Edgeworth's Taxation Paradox and the Nature of Demand and Supply Function, *Journal of Political Economy*, **40**, 577–616.

Howe, C. W. (1979) *Natural Resource Economics: Issues, Analysis and Policy* (New York: John Wiley & Sons).

Hultkrantz, L. (1982) *Skog för Nutid och Framtid* (Stockholm: EFI).

Hyde, W. F. (1980) *Timber Supply, Land Allocation and Economic Efficiency* (Baltimore: Johns Hopkins University Press).

Intriligator, M. D. (1971) *Mathematical Optimization and Economic Theory* (Englewood Cliffs: Prentice-Hall).

Jackson, D. H. (1980) *The Microeconomics of the Timber Industry* (Bowden: Westview Press).

Jevons, S. (1871) *The Theory of Political Economy* (London: Macmillan).

Johansson, P. O. (1981) An Outline of a Cost–Benefit Analysis of Increased Felling in Sweden, *Canadian Journal of Forest Research*, **11**, 637–42.

Johansson, P. O. (1982) Cost–Benefit Rules in General Disequilibrium, *Journal of Public Economics*, **18**, 121–37.

Johansson, P. O. (1983) On Cost–Benefit Rules in an Economy with Natural Resources, *Environment and Planning C: Government Policy*, **1**, 107–16.

Johansson, P. O. (1984a) Disequilibrium Cutting Rules, *Canadian Journal of Forest Research*, **14**, 321–5.

Johansson, P. O. (1984b) Disequilibrium Cost-Benefit Rules for Natural Resources, *Resources and Energy*, **6**, 355–72.

Johansson, P. O. (1984c) *On Socially Optimal Cutting Rules when Forest Lands are Managed Partly for Nontimber Purposes*, Swedish University of Agricultural Sciences, Department of Forest Economics, mimeographed. Forthcoming in *Canadian Journal of Forest Research*.

Johansson, P. O. and Löfgren, K. G. (1983a) Monopsony, Conjectural Equilibria and the Swedish Roundwood Market, *Forest Science*, **29**, 439–49

Johansson, P. O. and Löfgren, K. G. (1983b) The Timber Supply Function under Regulatory Constraints, *Canadian Journal of Forest Research*, **13**, 979–85.

Johansson, P. O. and Löfgren, K. G. (1984a) Six Different Results on the Properties of the Timber Supply Function, in Andersson, Å. E., Ohlsson, M. O. and Zachrisson, U. (eds) *Forest Sector Development. Issues and Analysis*, University of Umeå, Department of Regional Economics, report no. 1983:1.

Johansson, P. O. and Löfgren, K. G. (1984b) A Bargaining Approach to the Modeling of the Swedish Roundwood Market. Forthcoming in *Land Economics*.

Jungenfelt, K. G. (1973) *Mål och Model i Skogspolitiken* (Stockholm: Statens offentliga utredningar (Government Committee Report).

Just, R. E., Hueth, D. L. and Schmitz, A. (1982) *Applied Welfare Economics and Public Policy* (Englewood Cliffs: Prentice Hall).

Kaiser, F. H. and Norbury, F. L. (1984) *Multiple Use of Forest Resources from the Policy Perspective*, United States Department of Agriculture, Forest Service, Washington, DC, mimeographed.

Kallio, M. (1982) Sustained Yield in Forestry – Notes of Discussion, in Andersson,

Å. E. and Ohlsson, M. O. (eds) *Proceedings from a Nordic Workshop on Models for the Forest Sector*, University of Umeå, Department of Regional Economics, report no. 1982:1.

Kantor, B. (1979) Rational Expectations and Economic Thought, *Journal of Economic Literature*, **17**, 1422–41.

Katz, E. (1983) Relative Risk Aversion in Comparative Statics, *American Economic Review*, **73**, 452–3.

Katz, E., Paroush, J. and Kahana, N. (1982) Price Uncertainty and the Price Discriminating Firm in International Trade, *International Economic Review*, **23**, 389–400.

Katzner, D. W. (1970) *Static Demand Theory* (London: Macmillan).

Koopmans, T. C. (1965) On the Concept of Optimal Economic Growth, in *Pontificiae Academiae Scientiarium Scripta Varia*, vol. 28.

Kovenock, D. J. and Rothschild, M. (1983) Capital Gain Taxation in an Economy with an Austrian Sector, *Journal of Public Economics*, **21**, 215–56.

Kuuluvainen, J., Loikkanen, H. and Salo, J. (1983) The Timber Supply Behaviour of the Private Nonindustrial Forest Owners in Finland, *Metsäntutkimuslaitoksen Tiedonantoja*, **112** (in Finnish).

Lesourne, J. (1975) *Cost–Benefit Analysis and Economic Theory* (Amsterdam: North-Holland).

Lewis, T. R. (1976) Monopoly Exploitation of an Exhaustible Resource, *Journal of Environment Economics and Management*, **3**, 198–204.

Lippman, S. A. and McCall, J. J. (1980) *Progressive Taxation in Sequential Decision Making: Deterministic and Stochastic Analysis*, Western Management Science Institute, UCLA, working paper no. 304.

Löfgren, K. G. (1972) *Studier i Teorin för Prisdiskriminering*, Umeå Economic Studies, no. 2.

Löfgren, K. G. (1977) A Note on Output Under Discrimination, *Rivista Internazionale di Scienze Economiche e Commerciali*, **24**, 776–82.

Löfgren, K. G. (1983) The Faustmann-Ohlin Theorem: A Historical Note, *History of Political Economy*, **15**, 261–4.

Löfgren, K. G. (1984a) Endowments and Timber Supply, *European Review of Agricultural Economics*, **11**, 17–28.

Löfgren, K. G. (1984b) The Effect on the Socially Optimal Rotation Period in Forestry of Biotechnological Improvements of the Growth Function. Forthcoming in *Forest Ecology and Management*.

Löfgren, K. G. and Ranneby, B. (1983) *Two Papers on some Stochastic Aspects of the Supply and Demand for Timber*, Swedish University of Agricultural Sciences, Department of Forest Economics, report no. 46.

Löfgren, K. G. and Ranneby, B. (1984) *The Modeling of Behavioral Modes for a Firm Facing an Uncertain Supply Curve*, Swedish University of Agricultural Sciences, Department of Forest Economics, report no. 49.

Lohmander, P. (1984) *On the Optimal Investment Intensity under the Influence of Price Risk and Successive Decisions and on the Effects of Price Variations on the Expected Optimal Harvest Year*, Swedish University of Agricultural Sciences, Department of Forest Economics, working paper no. 29.

Lönnstedt, L. (1977) *Why does the Nonindustrial Forest Owner Cut?*, Swedish University of Agricultural Sciences, Department of Forest Economics, report no. 23.

Lutz, F. and Lutz, V. (1951) *The Theory of Investment of the Firm* (Princeton: Princeton University Press).

Lyon, K. S. and Sedjo, R. A. (1983) An Optimal Control Theory Model to Estimate the Regional Long-Term Supply of Timber, *Forest Science*, **29**, 798–812.

Malinvaud, E. (1953) Capital Accumulation and Efficient Allocation of Resources, *Econometrica*, **21**, 233–68.

Malinvaud, E. (1977) *The Theory of Unemployment Reconsidered* (Oxford: Basil Blackwell).

Malmborg, G. von (1969) *Ekonomisk Planering av Lantbruksföretaget* (Stockholm: College of Forestry).

Marchand, M., Mintz, J. and Pestieau, P. (1983) *Public Production and Shadow Pricing in a Model of Disequilibrium in Labor and Capital Markets*, Louvain La Neuve, CORE discussion paper no. 8315.

Marks, R. (1979) *Non-Renewable Resources and Disequilibrium Macrodynamics* (New York: Garland).

Masse, P. (1962) *Optimal Investment Decisions: Rules for Action and Criteria for Choice* (Englewood Cliffs: Prentice-Hall).

McDonald, J. M. and Solow, R. M. (1981) Wage Bargaining and Employment, *American Economic Review*, **71**, 896–908.

Mäler, K. G. (1974) *Environmental Economics: A Theoretical Inquiry* (Baltimore: Johns Hopkins University Press).

Miller, R. A. and Voltaire, K. (1983) A Stochastic Analysis of the Tree Paradigm, *Journal of Economic Dynamics and Control*, **6**, 371–86.

Mitra, T. and Wan, H. Y. (1982) *Some Theoretical Results on the Economics of Forestry*, Cornell University, Department of Economics, working paper no. 270.

Morey, E. R. (1984) Confuser Surplus, *American Economic Review*, **74**, 163–73.

Näslund, B. (1969) Optimal Rotation and Thinning, *Forest Science*, **15**, 446–51.

Nautiyal, J. C. and Fowler, K. S. (1980) Optimum in an Imperfect Stumpage Market, *Land Economics*, **56**, 213–26.

Nautiyal, J. C. and Rezende, J. L. (1983) Dynamic Investment Considerations in Forestry Project Evaluations, *Canadian Journal of Forest Research*, **13**, 640–7.

Neumann, J. von and Morgenstern, O. (1947) *Theory of Games and Economic Behavior*, 2nd edn (Princeton: Princeton University Press).

Ng, Y. K. (1979) *Welfare Economics. Introduction and Development of Basic Concepts* (London: Macmillan).

Nikaido, H. (1970) *Introduction to Sets and Mappings in Modern Economics* (Amsterdam: North Holland/American Elsevier).

Odhnoff, J. (1965) A Mathematical Note to Some Results by Tönu Puu, *Swedish Journal of Economics*, **67**, 80–2.

Ohlin, B. (1921) Till frågan om Skogarnas Omloppstid, *Ekonomisk Tidskrift*, **22**, 89–113.

Pindyck, R. S. and Rubinfeld, D. L. (1976) *Econometric Models and Economic Forecasts* (Tokyo: McGraw-Hill Kogakusha).

Pontryagin, L. S. (1962) *Ordinary Differential Equations* (London: Addison-Wesley).

Pontryagin, L. S., Boltyanski, V. S., Gamkrelidze, R. V. and Mishchenko, E. F. (1962) *The Mathematical Theory of Optimal Processes* (New York: Interscience Publishers, John Wiley & Sons).

Pratt, J. W. (1964) Risk Aversion in the Small and in the Large, *Econometrica*, **32**, 122–36.

Pressler, M. R. (1860) Aus der Holzzuwachlehre (zweiter Artikel), *Allgemeine Forst- und Jagdzeitung*, **36**, 173–91.

Puu, T. (1964) *Studier i det Optimala Tillgångsvalets Teori* (Uppsala: Almqvist & Wiksell).

Puu, T. (1967) Some Reflexions on the Theories of Choice Between Alternative Investment Opportunities, *Weltwirtschaftliches Archiv*, **99**, 107–25.

Roberts, K. W. S. (1982) Desirable Fiscal Policies under Keynesian Unemployment, *Oxford Economic Papers*, **34**, 1-22.

Robinson, V. L. (1974) An Econometric Model of Softwood Lumber and Stumpage Markets, 1946-1967, *Forest Science*, **20**, 171–9.

Ross, S. M. (1970) *Applied Probability Models with Optimization Applications* (San Francisco: Holden-Day).

Ruist, E. and Svennilson, I. (1948) Den Norrländska Skogsnáringens Konsunkturkånslighet under Mellankrigsperioden (Stockholm: Industriens Utredningsinstitut).

Salter, W. E. G. (1960) *Productivity and Technical Change* (Cambridge: Cambridge University Press).

Samuelson, P. A. (1947) *Foundations of Economic Analysis* (Cambridge: Harvard University Press).

Samuelson, P. A. (1976) Economics of Forestry in an Evolving Society, *Economic Inquiry*, **14**, 466–92.

Sandmo, A. (1971) On the Theory of the Competitive Firm under Price Uncertainty, *American Economic Review*, **61**, 65–73.

Sandmo, A. and Drèze, J. H. (1971) Discount Rates for Public Investments in Closed and Open Economies, *Economica*, **38**, 395–412.

Schaefer, M. B. (1954) *Some Aspects of the Dynamics of Populations Important to the Management of the Commercial Marine Fisheries*, Bulletin of the Inter-American Tropical Tuna Commission, 1.

Scott, A. D. (1955) The Fishery: The Objectives of Sole Ownership, *Journal of Political Economy*, **63**, 116–24.

Sedjo, R. A. (ed.) (1983) *Governmental Interventions, Social Needs, and the Management of US Forests* (Baltimore: Resources for the Future/Johns Hopkins University Press).

Seppälä, R. (1974a) Cutting Behavior of Private Forest Owners in Eastern Finland, *Folia Forestalia*, 189 (in Finnish).

Seppälä, R. (1974b) On the Supply of Roundwood in Finland, *Folia Forestalia*, 190 (in Finnish).

Shephard, R. (1953) *Cost and Production Functions* (Princeton: Princeton University Press).

Silberberg, E. (1972) Duality and the Many Consumer's Surpluses, *American Economic Review*, **62**, 942–52.

Smith, B. and Stephen, F. H. (1975) Cost–Benefit Analysis and Compensation Criteria: A Note, *Economic Journal*, **85**, 902–5.

Solow, R. M. (1963) *Capital Theory and the Rate of Return* (Amsterdam: North Holland).

Starrett, D. (1979) Second Best Welfare Economics in the Mixed Economy, *Journal of Public Economics*, **12**, 329–49.

Statistical Yearbook of Forestry 1953–1982: Official Statistics of Sweden, National Board of Forestry, Jönköping.

Streyffert, T. (1938) *Den Skogsekonomiska Teorin* (Oskarshamn: Skogsvårdsföreningens förlag).

Stridsberg, E. and Mattsson, L. (1980) *Skogen Genom Tiderna* (Stockholm: LTs Förlag).

Sydsaeter, K. (1981) *Topics in Mathematical Analysis for Economists* (London: Academic Press).

Thünen, J. H. von (1863) *Der Isolierte Staat*, vol. 3, published posthumously in 1863 by H. Schumacher.

Varian, H. R. (1978) *Microeconomic Analysis* (New York: W. W. Norton).

Weisbrod, B. A. (1964) Collective-Consumption Services of Individualized-Consumption Goods, *Quarterly Journal of Economics*, **78**, 471–7.

Wicksell, K. (1901) *Föreläsningar i Nationalekonomi, Häfte 1* (Lund: Gleerups). English edn (1934) *Lectures on Political Economy, Vol. I: General Theory*, ed. L. Robbins (London: George Routledge & Sons).

Index